Physical and Mathematical Simulation of COREX Ironmaking Process

COREX 炼铁工艺的物理数学模拟

Heng Zhou　　Mingyin Kou
Shengli Wu　　Jianliang Zhang

（扫码看全书彩图）

北　京
冶金工业出版社
2022

内 容 提 要

本书共分为4篇15章,全面阐述了物理数学模型与模拟在COREX系统中的应用,主要内容包括研究背景、COREX 预还原竖炉的物理数学模拟、COREX 熔融气化炉的物理数学模拟、COREX 填充床中粉尘行为的模拟研究四个部分,系统介绍了中心供气COREX 竖炉的气固流动、排料螺旋结构优化、上部调剂措施等内容,深入探讨了COREX 气化炉拱顶含碳颗粒燃烧及造气行为、循环煤气喷吹、物料下降行为以及布料方式对温度场影响等问题,详细研究了粉尘在填充床中的渗流扩散、沉积堵塞特性。本书为目前国内外关于 COREX 系统模型与模拟较新、较全面的著作,具有较高的参考价值。

本书可供冶金工程等相关专业的研究人员和工程技术人员阅读,也可供高等院校冶金专业的师生参考。

图书在版编目(CIP)数据

COREX 炼铁工艺的物理数学模拟=Physical and Mathematical Simulation of COREX Ironmaking Process:英文/周恒等著. —北京:冶金工业出版社,2022.4
ISBN 978-7-5024-9108-6

Ⅰ.①C… Ⅱ.①周… Ⅲ.①COREX 工艺—物理模拟—数学模拟—英文 Ⅳ.①TF557

中国版本图书馆 CIP 数据核字(2022)第 052128 号

Physical and Mathematical Simulation of COREX Ironmaking Process

出版发行	冶金工业出版社	电　话	(010)64027926
地　址	北京市东城区嵩祝院北巷39号	邮　编	100009
网　址	www.mip1953.com	电子信箱	service@ mip1953.com

责任编辑　杨　敏　任咏玉　美术编辑　燕展疆　版式设计　郑小利
责任校对　郑　娟　责任印制　李玉山

三河市双峰印刷装订有限公司印刷
2022年4月第1版,2022年4月第1次印刷
710mm×1000mm 1/16;17.25 印张;331 千字;261 页
定价 96.00 元

投稿电话　(010)64027932　投稿信箱　tougao@ cnmip.com.cn
营销中心电话　(010)64044283
冶金工业出版社天猫旗舰店　yjgycbs.tmall.com
(本书如有印装质量问题,本社营销中心负责退换)

Preface

Driven by the carbon emission mitigation and the fourth wave of industrial revolution, the iron and steel science and technology are undergoing green and intelligent transformation. The digital simulation and analysis of metallurgical reactors is a key foundation and important means to achieve intelligent manufacturing and the basic core of steel industry 4.0. On the other hand, it is one of the important means to control and optimize the reactors, and realize energy conservation, CO_2 emission reduce and green development.

COREX is the first industrialized smelting reduction ironmaking process. Due to the large-scale of COREX process, many issues have arisen, such as insufficient center gas distribution, high fuel ratio and so on. To solve these problems, some new technologies are applied in COREX-3000 process. For example, the center gas supply device in shaft furnace, the pulverized coal injection in the dome of melter gasifier, top gas recycling injection in COREX are applied. In this context, the application of physical and mathematical simulation of COREX is conducive to the in-depth understanding of COREX and provides a theoretical basis of the optimal operation parameters and the reasonable design of reactors. Based on the authors' research, this book comprehensively introduces the physical and mathematical simulation of COREX ironmaking process, including research background, simulation in COREX shaft furnace, simulation in COREX melter gasifier, and the fine particles behavior in the packed bed in COREX. It's hoped that the publication of this book can further strengthen readers' understanding of physical and numerical simulation methods, promote the development of COREX ironmaking process, and promote the green and intelligent development of ironmaking technology.

This book has 4 parts and 15 chapters. The first research background mainly introduces the development status of COREX process. The second part is the physical and numerical study of COREX shaft furnace, which mainly includes physical and DEM simulation of solid flow, the influence of discharging screw on solid flow, CFD-DEM simulate the gas-solid flow, and the inner characteristics in shaft furnace through CFD. The third part is the physical and numerical study of COREX melter gasifier. The main contents include the combustion characteristics in the dome zone of the melter gasifier, numerical simulation of pulverized coal injection in dome zone, mathematical study the top gas recycling into melter gasifier, solid flow in melter gasifier, and the physical simulate the influence of burden distribution on temperature distribution. The last part is the simulation of fine particles behavior in the packed bed of COREX, mainly including the physical and numerical simulation of dust accumulation in bustle pipe of COREX shaft furnace, the fine particles percolation behavior, the blockage behavior of fine particles in packed bed, and the gas-fine flow in a packed bed by CFD-DEM.

Dr. Heng Zhou, A. Prof. Mingyin Kou, Prof. Shengli Wu and Prof. Jianliang Zhang from University of Science and Technology Beijing are responsible for writing, and overall revision of this book. The chapters 2, 5, 10, 11 and 15 are written by Dr. Heng Zhou and prof. shengli Wu, chapters 3, 4, 6, 7, 9 and 13 were written by A. Prof. Mingyin Kou and Prof. Shengli Wu, and chapters 1, 8, 12 and 14 were written by Dr. Zhou Heng and Prof. Jianliang Zhang. The graduate students shun Yao, Yifan Hu, Bin Li, Zhong Zhang, Xu Tian and Wang Zeng participated in part of the writing of some chapters. In addition, Prof. Zongshu Zou and A. Prof. Zhiguo Luo from Institute of Multiphase Transport and Reaction Engineering, Northeastern University, participated in the writing of chapters 2, 4 and 12, and A. Prof. Lihao Han from Hebei Vocational University of Industry and Technology participated in the writing of chapters 1, 5, 6, 10, 11, 12 and 14.

The publication of this book has received the full support from leaders and colleagues of the School of Metallurgical and Ecological Engineering, and the State Key Laboratory of Advanced Metallurgy, University of Science and Technology Beijing. The authors would like to thank the National Key R&D Program of China (Grant Number: 2021YEC2902400, 2021YEC2902404), the National Natural Science Foundation of China (Grant Number: 51904023, 51804027), Fundamental Research Funds for the Central Universities (Grant Number: FRF-IP-20-04), the project of State Key Laboratory of Advanced Metallurgy (KF20-07, KF20-06) and the Fundamental Research Funds for the Central Universities and the Youth Techer International Exchange & Growth Program (Grant Number: QNXM20210011, QNXM20210009) for their financial supports. The authors would like to express our most sincere thanks. In addition, some previous works are referenced in this book and the authors express their sincere appreciation.

Authors
University of Science and Technology Beijing
2022. 01. 10

Contents

Part I Research Background

1 COREX Ironmaking Process ... 3

  1.1 Brief Introduction about Different Ironmaking Processes ... 3
    1.1.1 Blast Furnace ... 3
    1.1.2 Direct-reduced Ironmaking ... 5
    1.1.3 Smelting Reduction Ironmaking ... 6
  1.2 Brief Introduction about COREX Process ... 6
  1.3 COREX Process in China ... 8
  References ... 9

Part II Physical and Mathematical Simulation of COREX Shaft Furnace

2 Physical Simulation of Solid Flow in COREX Shaft Furnace ... 13

  2.1 Physical Modelling ... 14
    2.1.1 Apparatus ... 14
    2.1.2 Experimental Conditions ... 14
  2.2 Results and Discussion ... 16
    2.2.1 Characteristics of Solid Flow ... 16
    2.2.2 Effect of Variables on Solid Flow ... 18
    2.2.3 Effect of AGD Beams on Solid Flow ... 22
  2.3 Summary ... 25
  References ... 26

3 Mathematical Simulation of Solid Flow in COREX Shaft Furnace ... 27

  3.1 DEM Model ... 27
  3.2 Model Validity ... 29

3.3 Solid Flow Including Asymmetric Phenomena in Traditional SF 29
   3.3.1 Simulation Conditions 29
   3.3.2 Basic Solid Flow 31
   3.3.3 Effect of Discharging Rate 34
   3.3.4 Asymmetric Behavior of Solid Motion 36
3.4 Summary 39
References 39

4 Effect of Discharging Screw on Solid Flow in COREX Shaft Furnace 40

4.1 Influence of Screw Design 40
   4.1.1 Simulation Conditions 40
   4.1.2 Solid Flow in Base Case 43
   4.1.3 Effect of Screw Diameter 46
   4.1.4 The Optimized Case 48
4.2 Influence of Uneven Working of Screws 51
   4.2.1 Simulation Conditions 51
   4.2.2 Effect of Adjacent Inactive Discharging 51
   4.2.3 Effect of Separated Non-working Screws 56
   4.2.4 Effect of Discharge Rate 58
4.3 Summary 61
References 61

5 Gas-solid Flow in a Large-scale COREX Shaft Furnace with Center Gas Supply Device Through CFD-DEM Model 63

5.1 CFD-DEM Model 63
5.2 Model Validity 64
5.3 Influence of CGD on Gas-solid Flow 66
   5.3.1 Simulation Conditions 66
   5.3.2 Particle Velocity and Segregation 67
   5.3.3 Voidage and Gas Distribution 72
   5.3.4 RTD of Gas and Solid Phases 76
5.4 Influence of Burden Profile on Gas-solid Flow 81
   5.4.1 Simulation Conditions 81
   5.4.2 Burden Descending Velocity and Particle Segregation 82

    5.4.3 Gas Flow and Pressure Distribution 86
  5.5 Summary 88
  References 89

6 CFD Simulation of Inner Characteristics in COREX Shaft Furnace with Center Gas Distribution Device 91

  6.1 Mathematical Modelling 92
    6.1.1 Governing Equations 92
    6.1.2 Boundary Conditions 93
  6.2 Results and Discussion 95
    6.2.1 Model Validation 95
    6.2.2 Influence on Gas Flow 96
    6.2.3 Influence on Gas and Solid Composition 99
  6.3 Summary 103
  References 103

Part III Physical and Mathematical Simulation of COREX Melter Gasifier

7 Numerical Simulation of Combustion Characteristics in the Dome Zone of the COREX Melter Gasifier 107

  7.1 Mathematical Modelling 109
    7.1.1 Governing Equations 109
    7.1.2 Chemical Reaction Model 111
  7.2 Simulation Conditions 112
    7.2.1 Properties of the Recycling Dust 112
    7.2.2 Geometry and Boundary Conditions 112
  7.3 Result and Discussion 114
    7.3.1 Model Validation 114
    7.3.2 Interpretation of Base Model 116
    7.3.3 Effect of the Flow Rate of Rising Gas 121
    7.3.4 Effect of the Component of Rising Gas 123
    7.3.5 Effect of the Temperature of Rising Gas 124
  7.4 Summary 126
  References 126

8 Numerical Simulation of Pulverized Coal Injection in the Dome Zone of COREX Melter Gasifier ... 128

  8.1 Mathematical Model ... 129
    8.1.1 Governing Equations ... 129
    8.1.2 Chemical Reaction Model ... 130
  8.2 Simulation Conditions ... 132
  8.3 Results and Discussion ... 135
    8.3.1 Model Validation ... 135
    8.3.2 Effect of PCI in Dome Zone on the Performance of COREX MG ... 136
  8.4 Conclusions ... 144
  References ... 145

9 Mathematical Study the Top Gas Recycling into COREX Melter Gasifier ... 146

  9.1 Mathematical Modelling ... 146
    9.1.1 Description ... 146
    9.1.2 Establishment of the Mathematical Model ... 148
    9.1.3 The Top Gas Recycling Process ... 151
    9.1.4 Nitrogen Accumulation ... 151
    9.1.5 Calculation Method of CO_2 Emissions ... 153
  9.2 Results and Discussion ... 153
    9.2.1 Effect on Theoretical Combustion Temperature ... 155
    9.2.2 Effect on Dome Temperature ... 156
    9.2.3 Effect on Fuel Rate ... 157
    9.2.4 Effect on CO_2 Emissions ... 158
  9.3 Summary ... 160
  References ... 161

10 Influence of Cohesive Zone Shape on Solid Flow in COREX Melter Gasifier by Discrete Element Method ... 162

  10.1 Simulation Condition ... 162
  10.2 Results and Discussion ... 165
    10.2.1 Influence of Cohesive Zone Shape on the Mass Distribution ... 165
    10.2.2 Influence of Cohesive Zone Shape on the Velocity Distribution ... 166
    10.2.3 Influence of Cohesive Zone Shape on the Normal Force Distribution ... 168

  10.2.4 Influence of Cohesive Zone Shape on the Normal Force Distribution ··· 171
10.3 Conclusions ·············· 172
References ·············· 173

11 Influence of Burden Distribution on Temperature Distribution in COREX Melter Gasifier ·············· 174

11.1 Experimental ·············· 174
  11.1.1 Experimental Apparatus ·············· 174
  11.1.2 Experimental Conditions ·············· 175
  11.1.3 Experimental Procedures ·············· 176
11.2 Experimental Results and Discussion ·············· 176
  11.2.1 Influence of Radial Distribution of Relative DRI to Lump Coal and Coke Volume Ratio on the Temperature Distribution ·············· 176
  11.2.2 Influence of Coke Charging Location on the Temperature Distribution ·············· 178
  11.2.3 Influence of Coke Size on the Temperature Distribution ·············· 181
11.3 Conclusions ·············· 182
References ·············· 183

Part IV Simulation of Fine Particles Behavior in COREX

12 Experimental Study and Numerical Simulation of Dust Accumulation in Bustle Pipe of COREX Shaft Furnace with Areal Gas Distribution Beams ·············· 187

12.1 Experimental ·············· 187
12.2 Mathematical Model ·············· 189
12.3 Results and Discussion ·············· 190
  12.3.1 Characteristics of Dust Accumulation ·············· 190
  12.3.2 Effect of Blast Volume ·············· 193
  12.3.3 The Mechanism of Dust Accumulation ·············· 196
12.4 Conclusions ·············· 198
References ·············· 199

13 Numerical Study of Fine Particle Percolation in a Packed Bed ·············· 200

13.1 DEM Model ·············· 201

13.2　Simulation Conditions ……… 203
13.3　Results and Discussion ……… 204
　13.3.1　Comparison between Cubical and Sphere Particles ……… 204
　13.3.2　Effect of Cohesive Force on Percolation Behavior ……… 209
　13.3.3　Effect of Key Variables on Percolation Behavior ……… 214
13.4　Summary ……… 220
References ……… 220

14　Dynamic Analysis of Blockage Behavior of Fine Particles in a Packed Bed ……… 222

14.1　Blockage Behavior of Fine Particles ……… 222
　14.1.1　Simulation Conditions ……… 222
　14.1.2　Blockage Distribution and Mechanism ……… 223
　14.1.3　Effect of Charging Number of Fine Particles ……… 227
　14.1.4　Effect of Initial Velocity ……… 229
14.2　Influence of Cohesive Force ……… 231
　14.2.1　Simulation Conditions ……… 231
　14.2.2　Blockage Formation and Mechanism ……… 233
　14.2.3　Effect of Sticking Force on Blockage ……… 236
　14.2.4　Effect of Other Key Variables on Passage and Blockage Behaviour ……… 237
14.3　Summary ……… 240
References ……… 240

15　CFD-DEM Study of Fine Particles Behaviors in a Packed Bed with Lateral Injection ……… 241

15.1　Simulation Conditions ……… 241
15.2　Model Validity ……… 242
15.3　Results and Discussion ……… 243
　15.3.1　Effect of Gas Velocity ……… 243
　15.3.2　Effect of Diameter Ratio ……… 248
　15.3.3　Effect of Mass Flux ……… 252
　15.3.4　Effect of Rolling Friction ……… 255
　15.3.5　Clogging Mechanism ……… 258
15.4　Summary ……… 260
References ……… 261

Part I

Research Background

1 COREX Ironmaking Process

Iron-making industry is one of the most prosperous industries in the world and its related technologies have been drastically changed over decades. Generally, iron-making technologies include three process routes: blast furnace, direct reduction process and smelting reduction process. Although the latter two are shown to have numerous advantages in recent times, neither can replace the role of blast furnace as the mainstream iron-making process. The main reason is that direct reduction and smelting reduction technologies are still at infancy phase while the technology of blast furnace has reached a mature stage after hundreds of years' development and improvement. In the following paragraphs, each of the iron-making technologies will be briefly discussed.

1.1 Brief Introduction about Different Ironmaking Processes

1.1.1 Blast Furnace

The first blast furnaces appeared in the 14th century but the process inside the blast furnace remains the same, even though the blast furnace technology has been improved and evolved over decades. The modern blast furnace is a vertical counter-current heat exchange and chemical reactor for producing hot metal. Solid iron oxide burden is charged from top along with coke and flux, and as it descends in the furnace it is heated up by the ascending gas and the iron oxides are reduced into hot metal by the reducing gas. Fig. 1-1(a) shows the cross section of a typical blast furnace along with the inputs and outputs.

The blast furnace parts may be classified depending on the shape of the region (Fig. 1-1(a)). The upper cylindrical part of the furnace is known as the throat and is protected by refractory brick. Below the throat, there is the region with increasing diameter known as the shaft which extends to a cylindrical section or belly. After the belly, the diameter decreases again in the bosh region, where the blast enters the furnace. The bottommost portion of the furnace is called the hearth where the molten hot metal and the slag accumulate within a coke bed. The inner volume of the blast furnace is also classified into different zones (Fig. 1-1(b)) depending on the physical state of the burden and the chemical reactions occurring. The uppermost part of the furnace constitutes of the lumpy

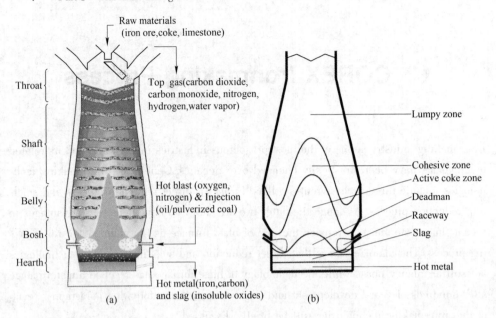

Fig. 1-1 Blast furnace diagram

(a) Cross-section of a typical blast furnace, classification based on shape of the furnace region;

(b) Different zones of the blast furnace classified on the basis of internal state

zone, where the burden remains solid. The iron ore, usually charged as haematite (Fe_2O_3) is first converted to magnetite (Fe_3O_4) and eventually to wustite1 (FeO) by the ascending reducing gas containing carbon monoxide (CO) which produces carbon dioxide (CO_2). Similar reduction reactions, but to a lesser extent, occur with hydrogen, forming water vapor. The temperature of the burden increases from the ambient to a constant temperature (900 ~ 1000 ℃) where both the burden and the gas attain nearly the same temperature. This region is called the thermal reserve zone. By contrast, the temperature of the gas decreases as it rises in the furnace and exits the top at 100 ~ 250 ℃. After the end of the thermal reserve zone, the wustite is reduced into iron (Fe).

A large number of other reactions take place in this region, including reduction of the other metallic oxides in the iron ore and the formation of slag. As some wustite always remains unreduced and the burden reaches higher temperatures than 1000 ℃, the Boudouard reaction ($C+CO_2 =\!\!= 2CO$), which is highly endothermic, will occur simultaneously. The net reduction reaction is $C+FeO =\!\!= Fe+CO$. Subsequently, the iron-bearing burden begins to soften and melt as the cohesive zone starts. This zone has alternate layers of highly pervious coke and semi-pervious iron-slag mix. The pervious coke layers or slits help the gas enter from the lower parts of the furnace to rise up towards the top. Therefore, an adequate size of the coke slits is very important for achieving a smooth furnace operation. At the lower end of the cohesive zone, the iron melts and percolates through the bed of solid coke. The upper part of the coke region is called the active coke zone. Here the coke is constantly replenished from the burden, as it slides to the combustion regions near the tuyeres known as the raceways. In the raceway, the coke is combusted to carbon monoxide by the incoming blast which consists of oxygen and (practically inert) nitrogen.

At the core of the bosh region lies a closely packed column of coke which does not react rapidly and is called the deadman. It provides support to the layered structures above. The hearth has a pool of liquid iron called hot metal with slag floating on top of it. The hot metal and the slag are tapped at regular intervals. The hot metal flows through a runner into a ladle or torpedo, which is transported to the steel mill for further processing. The slag, which is separated by gravity, is usually tapped into a slag pit or directly cooled and granulated. It is often sold as a by-product, e.g. to the cement or brick industries.

1.1.2 Direct-reduced Ironmaking

Direct reduction is the process that reduces iron ore (lumps, pellets or fines) in solid form to direct-reduced iron (DRI) without reaching the melting point of the iron ore. MIDREX Direct Reduction Plants produced over 64% of the world total production of direct-reduced iron, followed by HYL/Energiron. Both use pellets and lump ore as iron ore feed materials. FINMET and Circored are two natural gas based processes, and can directly use fine ore by the fluidized bed technology at the reduction stage.

Compared with a blast furnace, the direct reduction process is more energy efficient as it operates at a lower temperature, around 1000℃, which is 500℃ lower than that in the blast furnace. However, the quality of the reduced iron is not as desirable as pig iron from a blast furnace, as oxygen and silica are contained in the reduced products which need to be removed in subsequent operations at some added costs. For instance, the metallic iron content in direct-reduced iron is 80% to 88% compared to 95% in blast furnace iron.

In recent years, the global output of DRI had increased rapidly and exceeded 100Mt for the first time in 2018. The top five countries were India, Iran, Russia, Saudi Arabia and Mexico, with DRI output of 28.11, 25.75, 7.9, 6 and 5.97Mt, respectively. At present, dozens of direct reduction processes have been industrialized in the world, of which the gas-based shaft furnace process takes the leading position.

1.1.3 Smelting Reduction Ironmaking

The smelting-reduction processes have been developed as an alternative to the blast furnace process to overcome the disadvantages inherent with blast furnace process. These processes include pig iron production without the utilization of coke. The incentives behind the development of direct smelting-reduction processes can be summarized as follows.

(1) Utilization of a widely available variety of non-coking coals as reducing (carburizing agents and energy instead of coke).

(2) Utilization of a wider range of iron oxide feed stocks, which are harder to beneficiate in the blast furnace (i.e., waste oxides and high-phosphorus, sulfur-or titanium oxide-containing ores).

(3) Elimination of environmental emissions caused by the coke ovens, sinter plants, and induration kilns.

(4) Reduction of capital and operational costs by minimal ancillary plant and material-handling requirements.

(5) Lowered iron oxide feed stocks handling and preparation requirements. Preferably, direct utilization of iron ore fines or a concentrate rather than pellets, sinters, and lump ore.

(6) Economic operation at modest throughput rates as low as 1000 tons of pig iron per day.

(7) Compatibility of the product with the existing steel works.

(8) Flexible process that can be shut down and restarted easily.

(9) Elimination of energy losses by utilizing secondary energies within the process.

(10) Overall lowered energy consumption per ton of pig iron production since the blast furnace process is the most energy-intensive step in steel production.

Because of the above advantages, many such processes have been tested in pilot/demonstration plant scale such as AISI, AUSMELT, CCF, DIOS, FINEX, HISMELT, ROMELT and TECNORED. So far, HISMELT, COREX and FINEX are three smelting reduction technologies that operate on a commercial scale. These three processes will be discussed below.

1.2 Brief Introduction about COREX Process

COREX is the first smelting reduction process that operates at a commercial scale (Fig. 1-2). This process was developed by Voest-Alpine Industrieanlagenbau (VAI) and Deutsche Voest-Alpine Industrieanlagenbau (DVAI-Germany).

The first commercial Corex plant was located in Pretoria, South Africa, and was owned by ISCOR, with an annual capacity of 315000 tons of pig iron when it started operation in 1989. The largest COREX module, the COREX-3000, has an annual capacity of 1.5 million tons of iron. One of these modules was built at the Shanghai, China plant of Baosteel in 2007. A second module of the same size is expected to begin operation at the same plant in 2011. The smaller COREX-2000 module is in use at the plant of Jindal South West Steel Ltd in India, while another COREX-2000 module is in operation together with a Midrex module at the plant operated by ArcelorMittal in Saldanha, South Africa. The combined use of COREX modules and blast furnaces is allowing the Jindal company to make use of more than 70% of all of the blast-furnace sludge, limestone and dolomite screenings, and converter slag formed at the plant. These materials are used either directly or through the sinter or pellet plants, thus making it possible to minimize iron production costs. In addition, the export gas of the COREX module is being used to heat the blast in the blast-furnace stoves, in boilers, and at the sinter and pellet plants.

As shown in Fig. 1-2, the COREX is a two-stage process consisting of the reduction process in the reduction shaft and the smelting process in the Melter Gasifier. In the first reduction stage, lump ore, pellet and sinter are charged into the reduction shaft to produce the direct-reduced iron (DRI) by a reduction gas moving in counter flow. Through a discharge screw, the DRI could be conveyed from the reduction

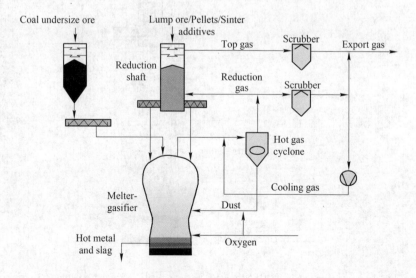

Fig. 1-2 Schematic diagram of the COREX process

shaft into the Melter Gasifier, where final reduction and melting take place. Coal in the Melter Gasifier could be combusted by oxygen injected into the Melter Gasifier, producing the reduction gas. This gas is cooled and is then blown into the reduction shaft to reduce the iron ores to DRI as described above. Lastly, hot metal and slag are discharged from the bottom of the Melter Gasifier as in the conventional blast furnace practice.

1.3 COREX Process in China

In 2005, Baosteel introduced the COREX process to China. In November 2007, the world's first COREX-C3000 was put into production in Luojing, Shanghai. In March 2011, the second COREX-C3000 was put into operation. However, due to the serious financial losses, these two COREX-C3000 were forced to stop production in 2011 and 2012 one after another. In 2013, the first COREX-C3000 was relocated to Xinjiang Bayi Steel due to resources and costs advantages. In 2015, it was put into operation after technical improvements. Fig. 1-3 shows the flow chart of the COREX process in China.

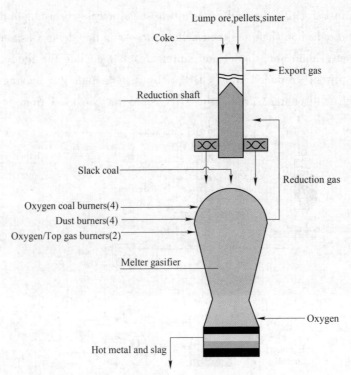

Fig. 1-3 Flow chart of COREX process in Bayi steel, Xinjiang province, China

Compared with the traditional COREX, some changes have taken place in the COREX in China. First, raw materials are changed from 20% lump ore+80% pellets to 55% sinter+40% pellets+5% lump ore. Second, fuels are changed from ump coal+metallurgical coke to metallurgical coke+nut coke+slack coal+PCI. Third, the burden distribution in shaft furnace has been modified to be more similar with the BF. Fourth, the dome in melter-gasifier has been equipped with four oxygen coal burners, four dust burners and two oxygen/top gas burners. The main economic and technical indexes are shown in Table 1-1. The cost of the COREX in Chian is approaching to BF. It is the largest and most advanced smelting reduction ironmaking furnace in the world at present.

Table 1-1 Main economic and technical indexes of COREX process in China

Item	Content	Units	2015	2017	2018
Key indexes	Smelting rate	t/h	140	150	160
	Metallization rate	%	25~35	45~55	40~55
	[Si]	%	1.0~2.0	1.14	0.6~1.0
Raw materials	Pellets	%	60	52	45
	Sinter	%	40	48	50
	Lump ore	%	0	0	5
Fuels	Coke	kg/t. HTM	306	250	200
	Nut coke	kg/t. HTM	120	120	130
	Slack coal	kg/t. HTM	354	314	300
	PCI	kg/t. HTM	0	195	200
	Fuel ratio	kg/t. HTM	780	879	830
Flux		kg/t. HTM	0	0	0
Productivity	Oxygen (standard state)	m³/t. HTM	550	480	480
	Operating rate	%			92~95
	Production	t/month			90000

References

[1] Zhang H M. CFD-DEM modeling of multiphase flow in a FINEX melter gasifier [J]. 2015, University of New South Wales, Sydney, Australia.

[2] Mitra T. Modeling of burden distribution in the blast furnace [M]. 2016, Åbo Akademi University, Turku, Finland.

[3] Wang Y J, Zuo H B, Zhao J. Recent progress and development of ironmaking in China as of

2019: an overview [J]. Ironmaking & Steelmaking, 2020, 47 (6): 640~649.
[4] Anameric B, Kawatra K S. Direct iron smelting reduction processes [J]. Mineral Processing and Extractive Metallurgy Review, 2008, 30 (1): 1~51.
[5] Kurunov I F. The direct production of iron and alternatives to the blast furnace in iron metallurgy for the 21st century [J]. Metallurgist, 2010, 54 (5-6): 335~342.

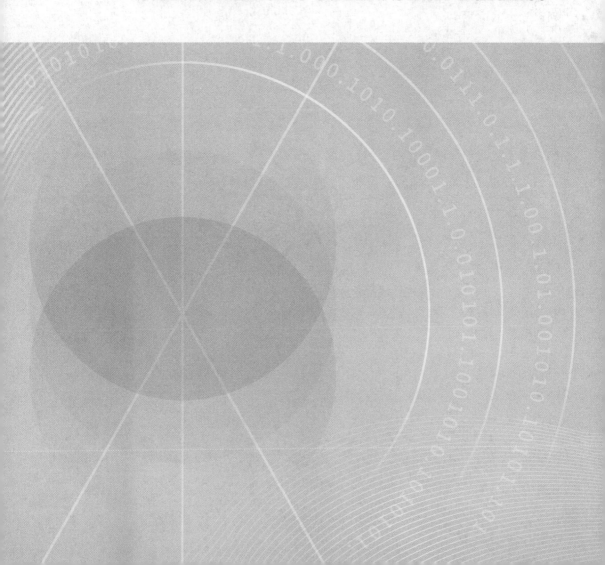

Part II

Physical and Mathematical Simulation of COREX Shaft Furnace

2 Physical Simulation of Solid Flow in COREX Shaft Furnace

COREX shaft furnace, which directly influences the quantity and quality indexes of the process, has been mainly used for producing directly iron (DRI) from lump iron ores and pellets. The shaft furnace (SF) reactor is a typical counter-current device in which reducing gases flow upward through the particle bed while the charged solid particles descend from the top and flow toward the bottom. The particles motion has significant effect on the flow of other phases and further on the smooth and efficient operation in these reactors. Therefore, the transient features of solids in COREX process is a hot research issue. COREX process realized its largest-scale production in terms of C-3000 in Baosteel at the end of 2007, and the second C-3000 module had begun operation at Baosteel since 2011. The shaft furnace of the second module is modified based on the first one and the most inspiring merit of this device is that a new technique called Areal Gas Distribution (AGD) is used. In the new design of COREX shaft furnace, two AGD beams are installed in the furnace to modify gas distribution with the purpose of improving iron ore reduction in the central part of the shaft furnace.

In the past, much research work has been carried out to understand the solid flow characteristics in blast furnace operated under different conditions. For the COREX shaft furnace, Lee investigated the solid flow profile and timeline of packed beds in shaft furnace with or without a guiding cone. Zhao simulated the solid flow in COREX shaft furnace through a scaled experimental model, the solid flow pattern, descending velocity are studied under various conditions. While these studies provide useful information, the furnace structure used in their work is a traditional furnace type without AGD beams.

In this chapter, solid flow behavior in COREX shaft furnace with AGD installed is experimentally studied using a semi-cylindrical model. Measurement of the solid flow timeline in a corn packed bed with screw dischargers is performed. The effects of a series of variables including gas flow rate, discharging rate and abnormal conditions are examined specifically. Evaluation of the effect of AGD beams on solid flow is also conducted.

2.1 Physical Modelling

2.1.1 Apparatus

The schematic diagram of COREX SF is shown in Fig. 2-1. The main difference between the traditional shaft furnace and new-design shaft furnace is that two AGD beams are installed just across the bustle. Fig. 2-1(c) shows a 1/20 scaled semi-cylindrical model that consists of three sections. The main section is the perspex vessel which is moulded to the shaft furnace shape. The lower section is the discharge system and the third section is the air supply system. The height of the model is 422mm in upper, 272mm in bustle and 292mm in low part. Air is blown into the vessel through the slots for the SF without AGD, but through the AGD inlets and slots for the new-design SF. When burden descend from top and flow through the beam, a gas free channel is formed under the AGD beam. So gas can flow through the AGD inlet into shaft centre via the channel. The schematic diagram explaining the gas flow from the bustle into furnace is shown in Fig. 2-2.

2.1.2 Experimental Conditions

As the physical properties, especially the repose angle of corn (diameter: 3mm, bulk density: 720kg/m^3, repose angle: 36°) is almost that of the average of coke (bulk density: 500kg/m^3, repose angle: 43.5°) and pellet (bulk density: 2000kg/m^3, repose angle: 32°), grains of corn were used to simulate the burden materials. The grains of corn were charged into the model from the top to maintain a certain height and extracted by five screw dischargers and the rate of corn particle extraction was controlled by induction motors. Air injected into the vessel from the bustle pipe, ascends through the apparatus and is exhausted from the top of the bed to the atmosphere. No air downflow occurred because the solid discharge was carried out within the closed box. The flow rate was controlled via a rotameter.

The Froude number relates the inertial forces acting on a phase to the gravity forces, which has a vital influence particularly on the descending behaviour of burden and flow behaviour of gas. Therefore, the solid descending velocity and blast volume in this work are determined based on the scale factor which is obtained by equalizing Froude number of the model and practical furnace. The Froude numbers are defined as follows, Fr_s for solid particles and Fr_g for gas.

$$Fr_s = \frac{U_s^2}{gl} \tag{2-1}$$

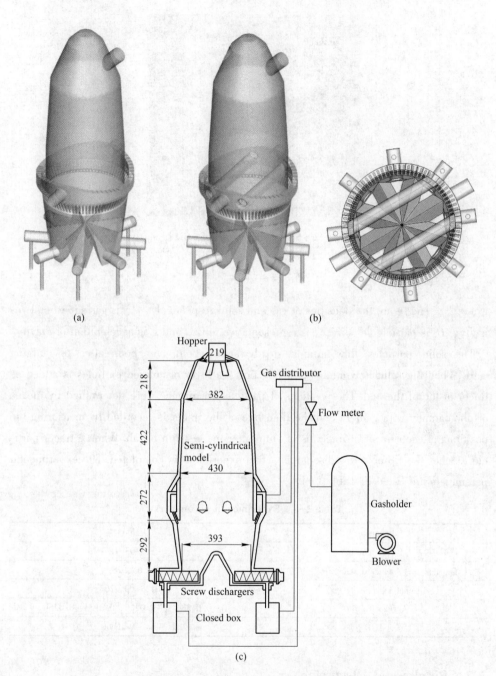

Fig. 2-1 Schematic diagram of COREX SF
(a) COREX SF without AGD; (b) Side and top views of COREX SF with AGD;
(c) The semi-cylindrical apparatus (mm)

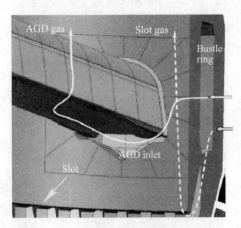

Fig. 2-2　Schematic diagram of AGD beam

$$Fr_g = \frac{\rho_g}{\rho_s - \rho_g} \cdot \frac{U_g^2}{gl} \qquad (2-2)$$

where, ρ_g and ρ_s are the densities of gas and solid respectively, U_g is gas superficial velocity, U_s is particle velocity, l is representative length and g is acceleration of gravity.

The solid particles flow attains steady state about one hour after discharging start. When the solid flow attains a steady flow, a layer of colored particles is added as tracer at top of the bed. The positions of particles in furnace were determined by means of photography with camera and the time line of the tracer is recorded by measuring the positions of each tracer at intervals of 1 min after the test run began. When a tracer particle reaches the tip of screw discharger, this experiment is completed. The experimental parameters are listed in Table 2-1.

Table 2-1　Experimental parameters

Parameter	Value
Bed height/m	1.03
Gas flow rate (standard state)/$m^3 \cdot h^{-1}$	0, 65, 78
Discharging rate/$kg \cdot s^{-1}$	0.0133, 0.0266
Asymmetric flow	0~0.0133, 0~0.0266, 0.0133~0.0266
AGD	With, Without

2.2　Results and Discussion

2.2.1　Characteristics of Solid Flow

Fig. 2-3 shows the internal flow structure for gas flow rate of $65 m^3/h$ (standard state), discharging rate of 0.0133kg/s in COREX SF. It can be observed that the whole flow area can be divided into four distinct flow zones. A man-made dead zone is set in lower central part to avoid a slow moving zone as used in plant operation. The plug flow zone typically occurs in upper part of the model. In this zone, particles descend with a relatively uniform velocity across the radius of the vessel, providing a good environment for uniform iron ore reduction resulted from uniform reaction time. The funnel flow zones occur directly above the tips of screw dischargers and these zones are characterized by high descending velocity. The quasi-stagnant zone described in this paper is the same as the previous papers, regarded as a sluggish descending zone. It is located in the lower part of the furnace between the funnel flow zone and furnace wall. The main reason for this phenomenon is that the discharging system in COREX furnace is screw and particles descend by gravity into the tips of the screw earlier. Then, the granules are horizontally transported to outlet through the spiral path between the blades by screw mechanical working. So the void near end screw is continuously filled by granules mostly horizontally transported from the vessel center. In this way, a quasi-stagnant zone is formed near the vessel wall and above the screw.

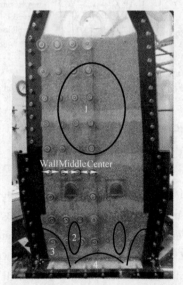

Fig. 2-3 Internal structure of solid flow in furnace
1—Plug flow; 2—Funnel flow; 3—Quasi-stagnant zone; 4—Guiding cone

The solid flow patterns for gas flow rate of $65m^3/h$ (standard state) and discharging rate of 0.0133kg/s are shown in Fig. 2-4. It can be seen that, in the plug flow zone of the upper part of the vessel, the descending of burden is uniform and a 'Flat' flow pattern is observed. This pattern continues till the bustle level. Then the flow pattern in bustle zone turns to 'Wave' shape just above the AGD beams. After passing through the AGD beams, a rudiment of 'W' shape profile appears. As descending and discharging proceeds, the solid flow pattern is gradually changed to a full 'W' shape in the bottom of the furnace.

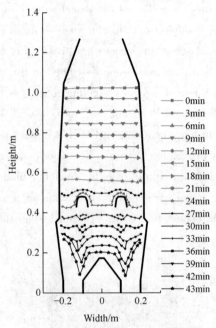

Fig. 2-4 Solid flow pattern of burden

2.2.2 Effect of Variables on Solid Flow

2.2.2.1 Gas Flow Rate

Fig. 2-5 shows the effect of blast volume on solid flow pattern. It was found that blast volume has little influence on solid particle behaviour and the flow patterns with different blast volumes are very similar. The solid flow profile turns out to be a Flat→Wave→W type. The influence of blast volume on solid flow in BF was investigated in previous work, indicating that the solid flow in the lower part of a BF is affected significantly by blast volume. The gas effect is reflected in the decrease of axial velocity of the material in the centre and causes an increase of descent speed near the walls, which was not observed in this study. This dissimilarity may be associated with the difference in the blast system. In a BF the tuyeres are located between bosh and hearth. The presence of gas causes an increase of the stagnant zone and restricts the main flow channel. For COREX, gas is injected through slots and AGD inlets in the bustle pipe zone located in the middle part of the shaft where it has little effect on the funnel flow zone and quasi-stagnant zone in lower part of furnace. Furthermore, differences in the main driving force for the solids flow also result in different flow behaviour. In previous work, coke combustion in the raceway is considered as the main driving force and the blast volume immediate impact the shape and size of the raceway, which significantly affect the solid flow. However, the main driving force for COREX is the mechanical working of the screw, therefore, the effect of blast volume on solid flow pattern is not the same as in a BF.

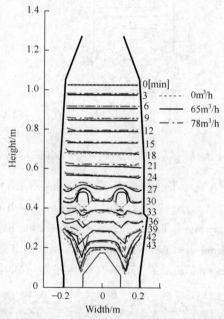

Fig. 2-5　Effect of blast volume on the solid flow pattern (Discharging rate=7.29r/min)

2.2.2.2　Discharging Rate

Fig. 2-6 shows the effect of discharging rate on solid flow pattern. The solid lines and broken lines in Fig. 2-6 represent timelines for discharging rates of 0.0133kg/s and 0.0266kg/s respectively. The numerical number corresponds to the elapsed time from charging. It can be observed that increasing the discharging rate has an effect on reducing the quasi-stagnant zone size. Comparing timelines of the two cases, within the same time, the distance between successive timelines of higher discharging rate is approximately two times of the lower case. This implies that a large solid flow rate can enhance the motion of particles and possibly reduce the period of static contacts between particles and the related clustering effect. But the effect solid flow rate on the utilization of reducing gas and chemical reactions needs further investigation.

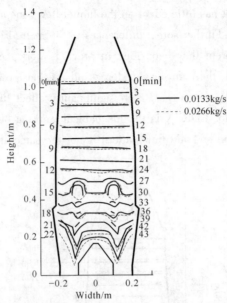

Fig. 2-6 Effect of discharging rate (gas flow rate under standard state = 65m³/h)

An interesting feature is revealed when comparing the AGD channel's cross section of the two cases, as shown in Fig. 2-7. The free channel for gas distribution under the AGD beam at higher discharging rate is larger than that at lower discharging rate and therefore, the triangle shaped free area increases with the increase of the discharging rate. The difference between the cross action of AGD channel with different discharging rate is 26.3%.

Fig. 2-7 Cross section of AGD channel at different discharging rates
(gas flow rate under standard state = 65m³/h)
(a) 0.0133kg/s; (b) 0.0266kg/s

2.2.2.3 Asymmetric Flow Patterns

Asymmetric flow behavior is the condition where uneven flow occurs within the vessel. In COREX SF, asymmetric flow can be caused by uneven working of screw dischargers around the SF circumference related to maintenance outage of screw or burden sticking above the screw. Uneven screw working causes different solid flow conditions within COREX SF and represents a serious operational problem. In this work, the asymmetric flow in COREX SF is simulated in the semi-cylindrical model furnace.

Fig. 2-8 shows the observed flow patterns under two different solid flow rates. Under extreme conditions that no solids are discharged from one side outlets, the flow patterns are significantly changed. There exists a very big stagnant zone on the left side for both cases. The particle layer from the top is distorted and the particles in left side of the vessel have a higher descending speed than that directly above the outlet. The same phenomenon was also observed in a previous work. However, the descending speed of particles at 0.82m height level of the model varies greatly across the width and the greatest descending speed occurs directly above the outlets, while on the opposite side of discharging outlets, descending speed is close to zero. This implies that the peak of the stagnant zone occurs close to this point in the model. Also, as observed in this asymmetric experiment, the size of stagnant zone is almost similar for different discharging rates.

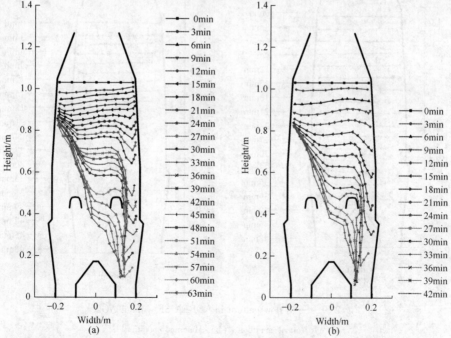

Fig. 2-8 Non-uniform descending of solids at asymmetric discharging
(a) 0~0.0133kg/s; (b) 0~0.0266kg/s

2.2.3 Effect of AGD Beams on Solid Flow

The effect of AGD beams on solid flow is studied as follows. The solid flow profiles of packed bed with and without AGD are measured under different operational parameters, including normal conditions and abnormal conditions.

Fig. 2-9(a) presents the solid flow patterns in COREX SF without AGD under gas flow rate of $65 m^3/h$ (standard state) and discharging rate of 0.0133kg/s. It can be seen that uniform descending occurred in the upper shaft and bustle zone, 'U' shape profile below bustle and 'W' shape discharging through the screws. Regarding the flow patterns in the plug flow zone of the upper shaft, the flow of solids is almost not affected by AGD beams under the current physical conditions. But it is not the case in bustle zone. The AGD effect can be reflected by the decrease of the descending speed of the burden above of the AGD beams.

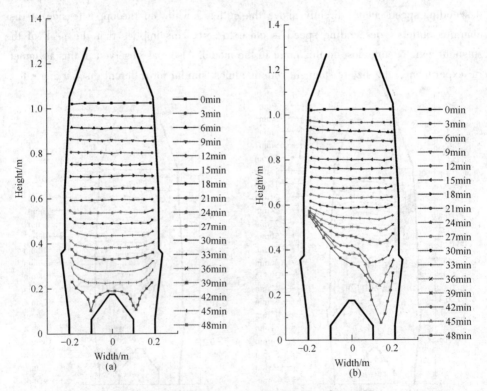

Fig. 2-9 Solid flow pattern in COREX SF without AGD
(a) Normal condition; (b) Abnormal condition

Burden descending speed in COREX SF with or without AGD for elapsed time of 24min to 33min under gas flow rate of $65m^3/h$ (standard state) and discharging rate 0.0133kg/s is shown in Fig. 2-10. It can be observed that burden descend in bustle zone for both cases during elapsed time from 24min to 33min (can be seen in Fig. 2-4 and Fig. 2-9(a)). Therefore, in this work, the solid descending speed in bustle zone is calculated from the timeline displacement divided by the elapsed time interval. In the case without AGD, the bustle zone experiences a relatively uniform descending through the cross-section of the vessel. On the other hand, in the case with AGD, above the AGD beams, the descending speed in the midway of vessel radius is lower than that in the center and near the wall. As also can be seen in Fig. 2-4, a smaller distance between successive timelines is observed, and the descending speed in center and near wall has a slight increase, but a gradual decrease in the middle. The main reason for this phenomenon is caused by AGD beams. The retardation of the flow area due to AGD beams installed in middle region of bustle causes an increase in descending speed along the centre and wall. Hence the solid flow profile in bustle zone turns to a 'Wave' shape just above the AGD beams. Below the AGD beams, a rapid burden descending channel is formed and particles in middle experience a rapid descending process. This is due to the void distribution. As burden descending from the top and flowing past the beams, an empty cavity (called as AGD channel in this work) is formed under the AGD beam, as shown in Fig. 2-7, and as expected, the void under the beam reaches maximum. So particles in middle descend by gravity into void quickly. In this way, a rudiment of 'W' shape profile appears below the AGD beams. Comparing with COREX SF without AGD, which has a flat profile in bustle zone, the solid flow pattern in new design furnace with AGD beams would experience 'Wave' and rudiment of 'W' shape in this region. Therefore this difference demonstrates that AGD beams influence the uniform descending in bustle zone to a certain extent.

Fig. 2-9(b) shows the observed flow patterns in COREX SF without AGD under abnormal conditions (0 ~ 0.0266kg/s). In the case, particles are removed from only one side of the vessel. Fig. 2-9(b) shows that uniform descending occurred in the upper part of the model furnace without AGD, the flow of solids is almost not affected by uneven working of screw dischargers. The descending velocity varies across the width at the height of 0.6m of the model. With the introduction of AGD beams, two main characteristics can be observed. The first one is that non-uniform descending is observed

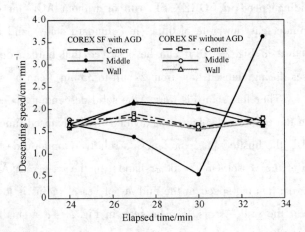

Fig. 2-10 The descending speed of solids at bustle level

in the upper part of the model. The second is that the peak of the stagnant zone is higher than that without AGD, which means that the active zone of particles decreases and the stagnant zone increases in size as the AGD beams installed. This implies that, under abnormal conditions, COREX SF with AGD could possibly increase the period of static contacts between particles and the related sticking effect. As a result, the choking of gas slots and the formation of cluster may be induced or enhanced. Therefore, it is significant to improve maintenance efficiency of screws, shorten the repair time and avoid the occurrence of asymmetric flow.

The effect of AGD on solid flow is further examined by observing the flow behavior of particles, as shown in Fig. 2-11. In both cases, burden is removed from one side at twice the rate of the other side. From Fig. 2-11(a), it can be observed that the descending in the upper region shows a 'Flat' pattern of horizontal layers similar to that of the bed with symmetric working of screws. This pattern continues to maintain till the bustle level. Then the discharging behavior in lower region turns to be affected by the asymmetric working of screw dischargers. This phenomenon demonstrates that abnormal operation of some screws has a minor effect on burden uniform descending in the upper shaft and bustle zone under the current condition. However, for the furnace with AGD, it can be seen that the flow behavior in one half of the model appears to be independent of the other half of the vessel. The AGD beams hinder the effect of mechanical working of screws in one side on the movement of solids in the other side. These variations demonstrate that the introduction of two AGD beams changes the solid flow significantly and causes the poor adaptability of SF to such abnormal operation.

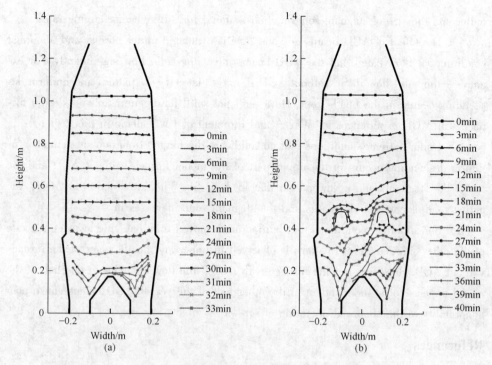

Fig. 2-11 Effect of AGD beams on solid flow pattern under asymmetric
discharging (0.0266 ~ 0.0133kg/s)
(a) Without AGD; (b) With AGD

2.3 Summary

An experimental study of solid flow behavior in COREX SF has been conducted under various conditions and evaluation of the effect of AGD (Areal Gas Distribution) on solid flow is investigated by comparing the solid flow patterns in SF with and without AGD. The findings can be summarized as follows:

(1) The internal flow structure in COREX SF can be divided into four distinct zones, including man-made deadman, plug flow, funnel flow and quasi-stagnant zone. The solid flow profile in SF with AGD shows a clear Flat→Wave→W type evolvement.

(2) Gas flow has a minor effect on burden descending behavior. Increasing the discharging rate has an effect of decreasing the quasi-stagnant zone size. A gas distribution channel is observed downstream of AGD beam and the triangle shaped free area increases with the increase of the burden discharging rate.

(3) For abnormal condition of asymmetric discharging, the flow pattern in SF with AGD shows that a big stagnant zone is formed on the opposite side of the discharging

outlet and the size of stagnant zone is almost similar for different discharging rates.

(4) The effect of AGD beams on solid flow is evaluated under normal and abnormal conditions. It is revealed that the AGD beams affect the solid flow significantly. For example, the solid flow in SF without AGD shows a Flat→U→W pattern and uniform descending occurs in the bustle zone. However, the solid flow pattern in new design furnace with AGD experiences a 'Wave' and rudiment of 'W' shape in this region.

(5) Under extreme conditions that no solids are discharged from one side outlets, uniform descending occurs in the upper part of the model furnace without AGD and the top of the stagnant zone is lower than that in AGD SF. With introduction of AGD, the active zone of particles decreases and the stagnant zone increases in size.

(6) For abnormal conditions that burden are removed from one side at twice the rate of the other side, a 'Flat' pattern is observed in upper and bustle zones of the furnace without AGD. However, for the furnace with AGD, the flow behavior in one half of the model appears to be independent of the other half of the vessel. AGD beams cause poor adaptability to the abnormal operation of screws.

References

[1] Ichida M, Nishihara K O, Tamura K, et al. Influence of ore/coke distribution on descending and melting behavior of burden in blast furnace [J]. ISIJ International, 1991, 31 (5): 505~514.

[2] Takahashi H, Komatsu N. Cold model study on burden behaviour in the lower part of blast furnace [J]. ISIJ International, 1993, 33 (6): 655~663.

[3] Takahashi H, Tanno M, Katayama J. Burden descending behaviour with renewal of deadman in a two dimensional cold model of blast furnace [J]. ISIJ International, 1996, 36 (11): 1354~1359.

[4] Wright B, Zulli P, Zhou Z Y, et al. Gas-solid flow in an ironmaking blast furnace-I: Physical modelling [J]. Powder Technology, 2011, 208 (1): 86~97.

[5] Lee Y J. A scaled model study on the solid flow in a shaft type furnace. Powder Technology [J]. 1999, 102 (2): 194~201.

[6] Lee Q, Zhao Y J, Zhang L J, et al. Simulation study on the solid flow of granular in shaft furnace of COREX [C] //China Metal Society. Proceeding of the fifteenth (2011) metallurgical reaction engineering conference. 2011.

[7] Shimizu M, Kimura Y, Isobe M, et al. Solids flow in a blast furnace under circumferential imbalance conditions. Tetsu to Hagane [J]. 1987, 73 (15): 194~201.

[8] Natsui S, Ueda S, Nogami H, et al. Analysis on non-uniform gas flow in blast furnace based on DEM-CFD combined model [J]. Steel Research International, 2011, 82 (8): 964~971.

[9] Li W G Operation status quo and technical problems of COREX-3000 [J]. Baosteel Technology, 2008, 6: 11~18.

3 Mathematical Simulation of Solid Flow in COREX Shaft Furnace

In the past, experimental investigation of the solid particles behavior in COREX shaft furnace (SF) with cold model has been carried out. However, the results obtained so far about SF solid flow are largely macroscopic, and the properties and microstructure at a particle scale for this particulate system are difficult to be obtained even with well controlled physical models. To overcome this problem, the discrete element method (DEM) has recently been used to study the solid descending behavior in COREX SF.

In this chapter, solid flow behavior in COREX SF is numerical studied by DEM. The simulation model is validated by comparing computed results with published experimental data. The discussion focuses on the effects of a series of variables including discharge rate and abnormal conditions on solid flow in COREX SF.

3.1 DEM Model

In DEM, every single particle in a considered system undergoes translational and rotational motion. The forces and torques considered include those originating from the particle's contacts with neighbouring particles, walls and surrounding fluids. The interaction force between two particles represented by spring-damper-friction plate is shown in Fig. 3-1. The particle i bears two kinds of force which are gravity $m_i g$ and the contact force between particle and particle, as well as particle and wall. In addition, the particle i bears two kinds of torque which are tangential torque and rolling friction torque. Based on the Newton's second law of motion, the governing equations for the translational and rotational motion of particle i can be written as

$$m_i(dv_i)/dt = \sum_{j=1}^{k} (F_{cn,ij} + F_{dn,ij} + F_{ct,ij} + F_{dt,ij}) + m_i g \qquad (3-1)$$

$$I_i(d\omega_i)/dt = \sum_{j=1}^{k} (T_{ij} + M_{ij}) \qquad (3-2)$$

where, m_i, I_i, v_i, ω_i represent mass, rotational inertia, translational velocity and rotational velocity of particle i, respectively. $m_i g$ represents the gravity of particle

i. $F_{cn,ij}$, $F_{ct,ij}$, $F_{dn,ij}$, $F_{dt,ij}$, T_{ij}, M_{ij} represent the normal and tangential contact forces, normal and tangential damp forces, tangential and rolling friction torques acting on i, respectively. k_i denotes the particle numbers contacting with particle i. Table 3-1 summarizes the equations used to calculate the forces and torques involved in Eqs. (3-1) and (3-2).

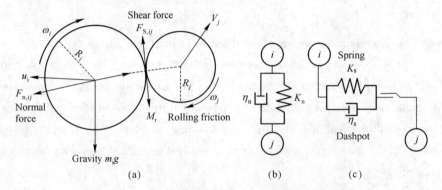

Fig. 3-1 Depiction of interaction forces between two particles

(a) Interactive forces between two particles; (b) Normal direction; (c) Shear direction

Table 3-1 Components of forces and torques acting on particle i

Force and torque		Symbol	Equation
Normal	Contact force	$F_{cn,ij}$	$-4/3 E^* \sqrt{R^*} \delta_n^{3/2} \hat{n}$
	Damping force	$F_{dn,ij}$	$-c_n (8 m_{ij} E^* \sqrt{R^* \delta_n})^{\frac{1}{2}} v_{n,ij}$
Tangential	Contact force	$F_{ct,ij}$	$-\mu_s |F_{cn,ij}| (1-(1-\delta_{t,ij}/\delta_{t,ij,max})^{\frac{3}{2}}) \hat{\delta}_t \cdots (\delta_{t,ij} < \delta_{t,ij,max})$
	Damping force	$F_{dt,ij}$	$-c_t \left(6\mu_s m_{ij} |F_{cn,ij}| \sqrt{1-\dfrac{\delta_{t,ij}}{\delta_{t,ij,max}}} \Big/ \delta_{t,ij,max} \right)^{\frac{1}{2}} v_{t,ij} \cdots (\delta_{t,ij} < \delta_{t,ij,max})$
Friction force		$F_{t,ij}$	$-\mu_s |F_{cn,ij}| \hat{\delta}_t \ (\delta_{t,ij} > \delta_{t,ij,max})$
Gravity		$F_{g,i}$	$m_i g$
Tangential torque		$T_{t,ij}$	$R_{ij} \times (F_{ct,ij} + F_{dt,ij})$
Rolling friction torque		$M_{r,ij}$	$\mu_{r,ij} |F_{cn,ij}| \hat{\omega}_{t,ij}^n$

where, $\dfrac{1}{R^*} = \dfrac{1}{|R_i|} + \dfrac{1}{|R_j|}$, $E^* = \dfrac{E}{2(1-v^2)}$, $\hat{n} = \dfrac{R_i}{|R_i|}$, $\hat{\omega}_{t,ij} = \dfrac{\omega_{t,ij}}{|\omega_{t,ij}|}$, $\hat{\delta}_t = \dfrac{\delta_t}{|\delta_t|}$, $\delta_{t,ij,max} = \mu_s \dfrac{2-v}{2(1-v)} \delta_n$, $v_{n,ij} = (v_{ij} \cdot \hat{n}) \hat{n}$, $v_{t,ij} = v_{ij} - v_{n,ij}$, $v_{ij} = v_j - v_i + \omega_j \times R_j - \omega_i \times R_i$

3.2 Model Validity

The validation of the present model is performed by comparing the predictions of the model with published experimental results of Lee. Fig. 3-2 shows a visual comparison of computed solid flow patterns in traditional SF with experimental result. One can see that there are good agreement both in size and shape of each zone, providing a solid support to the model. On this basis, the developed DEM model can be applied to the analysis of burden descending behaviours in latest design SF under various conditions.

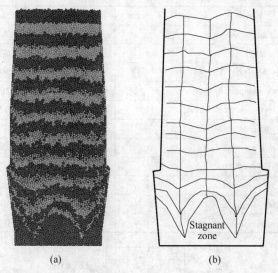

Fig. 3-2 Comparison between model prediction and experimental measurement
(a) Modelling; (b) Experimental

3.3 Solid Flow Including Asymmetric Phenomena in Traditional SF

3.3.1 Simulation Conditions

COREX SF is charged with a mixed burden of ore, pellet, coke, and flux. Because the quantity of pellets is much larger than the others, and the impact of particle properties such as density and diameter on solid flow is negligible, only the pellet was selected in this simulation. The number of burden particles in the COREX SF is much too large to be managed using the discrete element method as-is, so in this study, large particle size and scaled down model were used in order to model the process with DEM. Major dimensional parameters of the model SF are provided in Fig. 3-3 (at 1/20 of a practical COREX-3000 SF). The thickness of the model is a five-particle diameter ($5d_p$).

Simulation began with the random generation of a certain number of uniform spheres without overlaps, followed by a gravitational settling process for 3.5s. The particles were then discharged at a preset rate from the screw region. The burden surface decreased as burden discharged at screw outlets. When the burden surface reached a stock line level, burden was alternately charged. Material properties used in the simulation are listed in Table 3-2. The particle discharge rate was set very slow, only four particles every 100 time steps from the screw discharging region, equivalent to 0.251kg/s.

Table 3-2 Particle properties and simulation conditions

Variables	Value
Particle shape	Sphere
Particle number N	24800
Particle density $\rho_p / kg \cdot m^{-3}$	2500
Particle diameter d_p / mm	10
Sliding frictional coefficient μ_s	0.5
Rolling frictional coefficient μ_r	$0.005 d_p$
Young's modulus E/Pa	2160000
Poisson's ratio v_p	0.3
Time step $\Delta t/s$	1.0×10^{-4}
Discharge rate/kg · s^{-1}	0.251, 0.502, 0.753
Abnormal conditions/kg · s^{-1}	0~0.753, 0.502~0.753

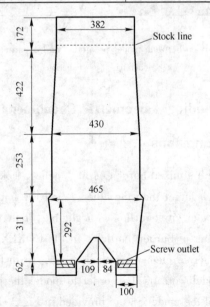

Fig. 3-3 Schematic diagram of calculation region of model COREX shaft furnace (Unit: mm)

3.3.2 Basic Solid Flow

The initial packed bed and snapshots of the flow patterns are shown in Fig. 3-4. The charging process was neglected, as it is simply not the focus of this study. Under the current simulation (discharge rate = 0.502kg/s), a relatively stable flow pattern was observed at 24.4s. The entire flow area was divided into four distinct flow zones, indicated schematically by the dashed lines in the snapshot at 34.6s. A man-made dead zone (Ⅳ) was set in the lower central part to avoid a slow moving zone, as used in plant operation. The plug flow zone (Ⅰ), which typically occurs in the upper part of the model, is where particles descend with a relatively uniform velocity across the radius of the model, providing a favorable environment for iron ore reduction. The funnel flow region (Ⅱ) occurs directly above the tips of solid outlet and is characterized by large descending velocity. The quasi-stagnant zone (Ⅲ), described in this paper similarly to previous reports, is regarded a sluggish descent zone. It located in the lower part of the furnace between the funnel flow zone and furnace wall.

Fig. 3-5 shows streamlines and timelines of tracer particles. Broken lines in Fig. 3-5 represent timelines from charging, and numerical number corresponds to the elapsed time from charging. At the top of the model furnace, in the plug flow zone, uniform descent occurred forming a flat pattern. This pattern continued down to the bustle zone. The flow pattern below the bustle zone then formed a 'U' shape. As descending and discharging continued, the solid flow pattern gradually changed into a 'W' shape in the bottom of the furnace. The crosses and open circles in the figure represent initial and disappearing positions of particles, respectively. The particles charged near-parallel to the wall descended linearly along the wall; particles charged in the middle, however, moved downward up to the guiding cone before approaching the outlets.

The geometric design of the screws in the bottom of the COREX SF directly affects solid flow patterns. In a previous study, the effects of screw flight diameter on particle descending velocity along the radius during the discharging process was studied, and found that particle descending velocity was the largest at the tips of screws and reduced from the center to wall areas when the first flight was large. Similar particle descent behavior was observed in this work. Optimizing the screw fight diameter near the center creates uniform descending velocity. Generally, increasing the pitch of the screws, decreasing the shaft diameter, increasing the screw diameter, or a combination of these can increase screw capacity in the direction of flow and achieve an even flow pattern. Detailed effects of screw geometric design on solid flow throughout the COREX SF are currently under research, specifically after matching the screws and inner structure of the furnace, including AGD (Areal Gas Distribution) beams installed in the bustle zone.

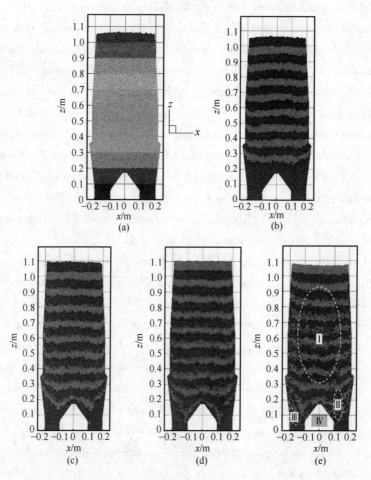

Fig. 3-4 Snapshots of solid flow pattern from initial to steady state
(a) $t=0s$; (b) $t=20s$; (c) $t=23.3s$; (d) $t=24.4s$; (e) $t=34.6s$

Thorough examination of the microscopic structures of solid flow, such as force structure, helps build a better understanding of complex solid flow behavior and underlying mechanisms. Fig. 3-6 shows a snapshot of normal contact force, where the length of arrows represents the magnitude of the force between two particles. Particles exhibited weak force network in the funnel flow region, because particles in this zone flowed rapidly and more voids and disconnections existed between them. Under a guiding cone setting in the lower central part of the furnace, strong normal force presented at the top of the man-made dead zone and extended to the bustle zone. In the COREX SF without a guiding cone, as shown in Fig. 3-7(b), the particles in the lower central bottom, instead, experienced strong normal force, because they needed to support the particles above them. A similar phenomenon was observed in a previous study by Zhou et al., who simulated the force structure of a model BF and found that under lower gas flow, strong force chains existed at the lower central part and extended to the top part of the furnace.

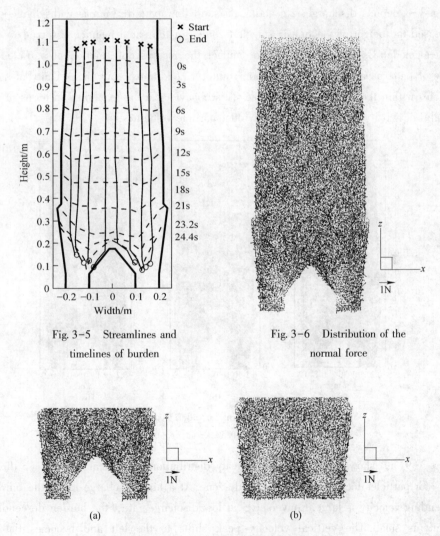

Fig. 3-5 Streamlines and timelines of burden

Fig. 3-6 Distribution of the normal force

Fig. 3-7 Normal stress distribution in the bottom of furnace
(a) With man-made guiding cone; (b) Without guiding cone

3.3.3 Effect of Discharging Rate

COREX SF is a typical gas-solid counter-current reactor. The effect of gas flow rate on solid flow was investigated in a previous experiment and in Hou's DEM model. According to these studies, gas flow rate has only minor effects on burden descending behavior; to this effect, gas flow is not included in the following model study, which instead focuses solely on discharge rate and abnormal conditions effects on solid flow.

Fig. 3-8 depicts discharge rate acting on solid flow pattern. Increased discharge rate decreased the height of the quasi-stagnant zone, in the same manner observed in the physical model. Discharge rate did not impact the macroscopic motion of particles or shape of patterns above the bustle under uniform conditions, but the velocity of solid particles within the furnace did increase. This behavior can be explained in terms of the probability density distributions of individual particle velocities.

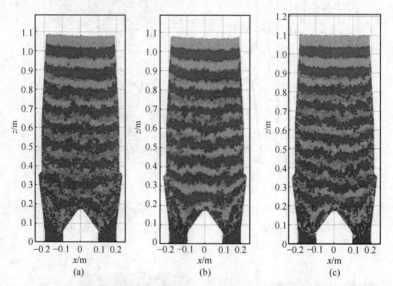

Fig. 3-8 Solid flow patterns for different discharge rates
(a) 0.251kg/s; (b) 0.502kg/s; (c) 0.753kg/s

Fig. 3-9(a) shows the probability density distribution of vertical velocity (z direction) for particles located below the bustle zone. At a high discharge rate, the burden descending velocity is large. Conversely, at low discharge rate, the burden descending velocity is small. The vertical velocity peak shifts to the left and becomes flat as discharge rate increases, suggesting that the number of motionless/inactive particles reduced, and that more particles became active with relatively high vertical velocities.

Fig. 3-9(b) shows the probability density distribution of horizontal particle velocity

(x direction) for different discharge rates. The curve shows Gaussian distribution, due to geometric symmetry. It becomes flatter as solid flow rate increases, indicating where more particles obtained higher horizontal velocity. This result is consistent with observed changes in the quasi-stagnant zone. As solid flow rate increased, the height of the quasi-stagnant zone reduced and more particles flowed toward the tip of the outlet at increased horizontal velocity. These phenomena are in general agreement with common sense, however, microscopic information obtained through DEM fosters a more comprehensive understanding of solid flow behavior inside the COREX SF.

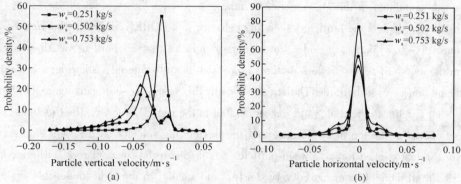

Fig. 3-9 In the lower part of COREX SF for different particle discharge rates
(a) Probability density distributions of particle vertical velocities; (b) Particle horizontal velocities

The probability density distribution of the normal contact force for different discharge rates is shown in Fig. 3-10. Discharge rate had rather little influence on the probability density distribution of normal contact force. Average normal contact force decreased slightlyas discharge rate increased, because under current discharge rate, most

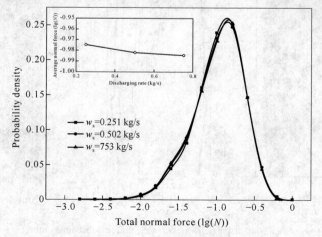

Fig. 3-10 Probability density distribution of the normal contact force

particles are in contact and resultant contact force can be spontaneously reconstructed.

3.3.4 Asymmetric Behavior of Solid Motion

Asymmetric flow behavior is characterized by uneven flow within a vessel. In a COREX SF, asymmetric flow is typically caused by imbalanced function of screw dischargers around the SF's circumference related to maintenance outage of screws or burden sticking above the screws. Uneven screw function causes different solid flow conditions within the COREX SF, and represents a serious operational problem. In this study, asymmetric flow was simulated in the DEM model for analysis.

Fig. 3-11 depicts the burden descending behavior in a COREX SF under abnormal conditions (0~0.753kg/s). Under the extreme condition that no burden was discharged from the left outlet, solid flow patterns were significantly altered, showing a sizeable stagnant zone on the left side. The stream lines in this case, as shown in Fig. 3-11(b), were not parallel to the wall, and instead shifted to the right. Moreover, the timelines became curved at the top of the furnace and the descending velocity on the right side grew faster. Normal contact force between particles was larger in the stagnant zone, where particles stayed motionless or moved very slowly, and smaller in the right funnel flow region where particles moved rapidly (Fig. 3-11(c)).

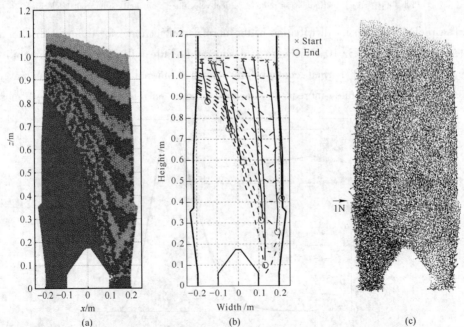

Fig. 3-11 Descent behavior of particles under non-uniform discharge rate (0~0.753kg/s)
(a) Solid flow patterns; (b) The stream lines and timelines; (c) Normal contact force

A comparison of probability density distribution of normal contact force between normal and abnormal conditions is shown in Fig. 3-12. The peak shifts to the right for abnormal conditions, reflecting stronger particle - particle contact. Average normal contact force increased under abnormal conditions, where a large stagnant zone formed on the opposite side of the working outlet and more particles remained in direct contact with strong interaction force. Normal contact force distribution along the vertical axis is shown in Fig. 3-13. Fig. 3-13(a) and Fig. 3-13(b) represent normal force distributions divided between the left and right part for uniform discharge and non-uniform discharge conditions. Under the symmetrical conditions shown in Fig. 3-13(a), contact force distributions for left and right parts were clearly quite similar. Normal force at the bottom part of the left side shown in Fig. 3 - 13 (b), however, was larger. Under extreme conditions, motionless particles located in the left stagnant zone showed potential to increase the period of static contact and resultant sticking effect. Choked gas slots and cluster formation are potentially induced or enhanced, as well. Improving the maintenance efficiency of screws, shortening repair time, and avoiding asymmetric flow are thus considered highly critical considerations as far as overall furnace function.

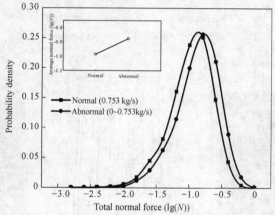

Fig. 3-12 Comparison of probability density distribution of normal force between normal and abnormal conditions

Asymmetric flow was further examined under the condition that particles were discharged from both outlets at different discharging rates, as shown in Fig. 3-14. Particles behaved differently on each side of the center line. On the right side, particles descended faster than those on the left due to increased discharge rate. The particle layer at the top was distorted due to the differing descending velocity on either side. Fig. 3-14(b) shows the resultant stream lines and timelines, clearly illustrating variations in particle descent behavior. The first and second tracer from the left wall show similar tendencies as those shown in Fig. 3-5, where linear descent was observed along the wall above the bustle, charged into the middle outlet below the bustle. The stream lines of the third and fourth tracer from the left wall were not parallel to the wall, and shifted to the right above the guiding cone. Slowed flow area caused by the guiding cone in turn sent the tracers toward the left outlet along the inclined surface of the guiding cone. Fig. 3-14(c) shows normal contact force, illustrating that the asymmetry of the force network reduced slightly compared to Fig. 3-11(c).

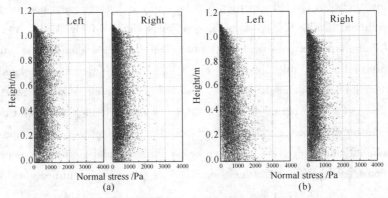

Fig. 3-13 Normal force distribution along vertical axis
(a) Uniform discharging; (b) Non-uniform discharging

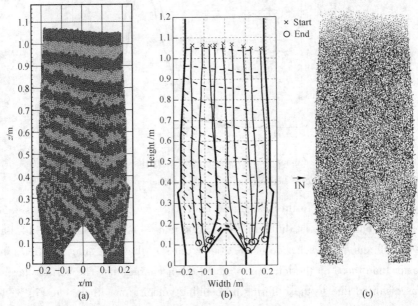

Fig. 3-14 Descent behavior of particles under non-uniform discharge rate (0.502~0.753kg/s)
(a) Solid flow patterns; (b) The stream lines and timelines; (c) Normal contact force

3.4 Summary

The discrete element method (DEM) was employed to investigate solid flow in a model COREX SF. The model was validated by comparing its solid flow patterns to those of a previous experiment. The effects of certain variables including discharge rate and abnormal conditions on solid flow, and the effect of AGD on burden descending behavior are simulated. The findings can be summarized as follows:

The solid flow in SF without AGD shows a Flat→U→W pattern and uniform descending occurs in the bustle zone. However, the solid flow pattern in SF with AGD experiences a 'Wave' and rudiments of 'W' shape in this region. AGD beams influence the uniform descending in bustle zone. Besides, the introduction of two beams increases the complexity of normal force distribution. The quite large normal force in local regions above beams aggravates clustering problem in bustle area.

Increasing the discharge rate decreased the quasi-stagnant zone size, but did not affect the macroscopic motion of particles or shape of patterns above the bustle. The velocity of solid particles within the furnace increased as discharge rate increased, while the average normal contact force decreased slightly.

Under an abnormal condition where no burden is discharged from the left outlet, resultant solid flow patterns changed significantly. A large stagnant zone formed opposite the working outlet, and motionless particles located in the left stagnant zone were able to potentially increase the period of static contacts and sticking effect. Under asymmetric condition, where particles are discharged from both outlets at differing discharge rate, resultant solid flow patterns were asymmetrical. Microscopic analysis confirmed that asymmetric conditions did not significantly affect normal contact force, but showed considerable impact on flow patterns.

References

[1] Lee Y J. A scaled model study on the solid flow in a shaft type furnace [J]. Powder Technology, 1999, 102 (2): 194~201.

[2] Cundall P A, Strack O D L. A discrete numerical model for granular assemblies [J]. Geotechnique, 1979, 29 (1): 47~65.

[3] Hou Q F, Samman M, Li J, et al. Modeling the gas-solid flow in the reduction shaft of COREX [J]. ISIJ international, 2014, 54 (8): 1772~1780.

[4] Wright B, Zulli P, Zhou Z Y, et al. Gas-solid flow in an ironmaking blast furnace—I: Physical modelling [J]. Powder Technology, 2011, 208 (1): 86~97.

4 Effect of Discharging Screw on Solid Flow in COREX Shaft Furnace

At the bottom of COREX shaft furnace (SF), eight screws are symmetrically located in the center of the furnace to discharge the burdens into the melter gasifier (MG) for further melting reduction. As an industrial screw feeder system, the condition of discharging screw can directly determine the burden descending behavior, which has a significant effect on the gas distribution, and further on the efficiency and effectiveness of process conducted in the SF. Hence, it is of great importance to investigate the effect of discharging screw on solid flow in COREX SF.

In this chapter, a three-dimensional actual size model of COREX SF was established based on DEM. Four types of burdens, including pellet, ore, flux and coke were considered in this model. In Section 4.1, the influence of screw design on burden descending velocity and particle segregation in SF was investigated. In Section 4.2, the effect of the uneven working of screw dischargers around the circumference of the shaft furnace on the solid flow was studied.

4.1 Influence of Screw Design

4.1.1 Simulation Conditions

The COREX SF geometry that was employed in this study is shown in Fig. 4-1. The simulation was begun with a random generation of a certain number of well-mixed particles to form a packed bed in the model SF. The eight screws were then started in rotation at a pre-set rate to extrude the particles. The parameters used in the present simulation are the same as shown in Table 4-1.

Table 4-1 Material properties and simulation parameters

Parameters	Pellet	Ore	Coke	Flux	Wall
Density ρ/kg·m^{-3}	3425	4760	1100	2800	7850
Shear modulus G/MPa	10	10	2.2	10	79000
Poisson's ratio v	0.25	0.21	0.22	0.21	0.30
Diameter d/cm	10	13	21	15	—

Continued Table 4-1

Parameters		Pellet	Ore	Coke	Flux	Wall
Mass ratio $w/\%$		51	34	7	8	—
Number of particles		440000	96052	20193	24860	—
Restitution coefficient e	Pellet	0.2	0.3	0.3	0.3	0.5
	Ore	—	0.5	0.3	0.3	0.5
	Coke	—	—	0.5	0.3	0.5
	Flux	—	—	—	0.3	0.5
Static friction μ_s	Pellet	0.5	0.5	0.5	0.5	0.4
	Ore	—	0.5	0.5	0.5	0.4
	Coke	—	—	0.6	0.5	0.4
	Flux	—	—	—	0.5	0.4
Rolling friction μ_r	Pellet	0.02	0.05	0.05	0.05	0.05
	Ore	—	0.05	0.05	0.05	0.05
	Coke	—	—	0.05	0.05	0.05
	Flux	—	—	—	0.05	0.05
Time step t/s		\multicolumn{5}{c}{5.8×10^{-4}}

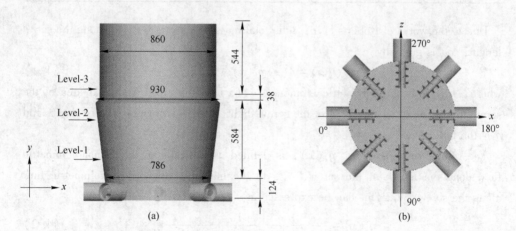

Fig. 4-1 The 3D model with screws used in the simulation (Units: cm)
(a) The side views; (b) Top views

There are numerous screw designs used in industry, which are based on the variation of blade diameter, screw core diameter and pitch angle. In this study, we investigate three typical designs with changing the blade diameter and one optimized screw. Each of the screws varied one or more design variables while keeping the others constant. Schematic diagram of the different flight diameter screws are shown in Fig. 4-2. All these screws are with constant screw core diameter D_c and screw length L, which are 52.6cm and 354cm. Detailed condition of screws is tabulated in Table 4-2, where the pitch for screw B and screw C are the same.

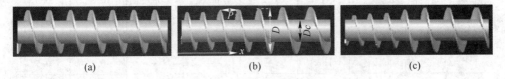

(a)　　　　　　　　　(b)　　　　　　　　　(c)

Fig. 4-2　Schematic diagram of screw
(a) Screw A; (b) Screw B; (c) Screw C

Table 4-2　Geometric properties of the screw designs

	Number	1	2	3	4	5	6	7
P/cm	Screw A	50	50	50	50	50	50	50
	Screw B and C	37.5	39.4	43	46	51.8	62.9	73.3
D/cm	Screw A	100	100	100	100	100	100	100
	Screw B	90	90	100	100	110	110	110
	Screw C	75	75	85	85	110	110	110

Due to the works of Roberts et al., the total mass flow rate, $Q(x)$, along the screw length, x, is given by

$$Q(x) = A(x)\eta_v(x)p(x)\omega\rho \tag{4-1}$$

where, $\eta_v(x)$ is the volumetric efficiency, $A(x)$ is the cross-sectional area of the screw flight, $p(x)$ is the pitch of the screw flight, ω is the angular screw velocity and ρ is the bulk density.

The volumetric efficiency, $\eta_v(x)$, is defined as the ratio of the actual volumetric flow along the length of the screw, V_L, to the maximum theoretical volumetric flow along the screw, V_{Lt}. This can be expressed as

$$\eta_v(x) = \frac{1}{\tan\alpha_m(x)\tan[\varphi_f + \alpha_m(x)] + 1} \tag{4-2}$$

where, φ_f is the screw face wall friction angle and $\alpha_m(x)$ is the mean helix angle of the flight along the screw length, x. The mean helix angle is given by

$$\alpha_m(x) = \tan^{-1}\left[\frac{p(x)}{\pi D_m}\right] \tag{4-3}$$

where, $p(x)$ is the screw pitch along the screw length and D_m is the effective mean screw diameter defined as the average of the outer screw D and inner core D_c diameters.

The withdrawn volume, $Q(x)/\omega\rho$, and the net increment of the volume, from No. 1 to No. 5 pitch (only five pitches are working in the SF, the other are in the casing.) for the screw A (refer to base case later), screw B and screw C are calculated and presented in Fig. 4-3. From Fig. 4-3(a), the withdrawn volume for the base case is constant along the screw length while it increases from No. 1 to No. 5 pitch for the screw B and screw C. The variation of the net increment of volume shows that almost all the burden transported by the screw has filled in the first flight for the base case. It may has a drawdown pattern that is far from uniform as is physically possible to obtain. The first flight fills only about 48.2% and 31.2% of the final screw transport volume for screw B and screw C, respectively. And the net increment of volume increase from No. 2 pitch along the screw for the both screws. This is significantly more even than for the base case. It is consistent with the intention of the screw design, which is to drawdown less material at the tip of the screw (due to its small blade diameter) and to increase the drawdown further along the screw because of the gradual tapering of the screw flight.

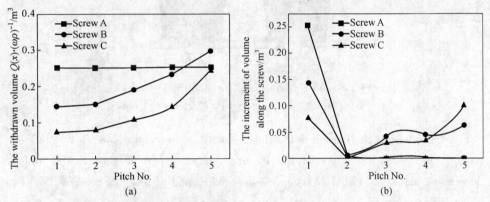

Fig. 4-3　Comparison of the different screws of the variation along the screw
(a) Withdrawn volume; (b) Net increment of volume

4.1.2　Solid Flow in Base Case

The screw used as the base case is the simplest screw in this work and one of the most common designs found in industry. The screw has a constant flight diameter D, constant core D_c and a constant pitch p. In Fig. 4-4, three horizontal layers of particles at a height of 2.5m height were selected and colored in orange to highlight their profile changes during discharging. Due to the complicated geometry of the SF and the design of the base screw, one main feature was observed in which the burden is primarily drawn down from the tip of the screw. The main reason is that the effect of the first flight is quite large and direct for the particle movement. As the screw flight rotates around its axis, an empty space is formed at the tip of the screw and then the particles are preferentially drawn down into the first flight. The similar phenomenon were also found in previous works with constant-pitch screw feeders. This is also consistent with the theoretical analysis of screw feed. In order to specifically characterize the internal flow in the SF, the burden descending velocity in SF will be discussed in detail in later.

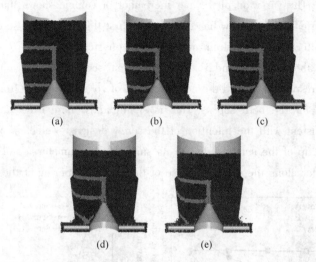

Fig. 4-4 Snapshots of cross-sections in the SF in the base case
(a) $t=0s$; (b) $t=2s$ (c) $t=4s$; (d) $t=6s$; (e) $t=8s$

Fig. 4-5 presents the burden descending velocity in the center plane (0° ~ 180° direction) of the SF. It can be seen that the upper region experiences a relatively uniform descending across the radial direction. As descending and discharging proceeds, a complicated flow zone occurs in the lower part of the SF. The largest descending velocities were observed above the tip of the screws. Sluggish moving particles typically locate

Fig. 4-5 Burden descending velocity in the center plane (0° ~ 180° direction) of the SF

above the end of the screw. This is because the withdrawn volume at each pitch is nearly constant for the constant-pitch screw. Since the withdrawn volume near screw end is filled by granules that are transported horizontally from the tips, no more voids for the particles descend into the end pitch of screw, so a lower velocity zone is observed.

For a universal understanding of the particle descending behavior in SF, three levels were examined to analyze the particle descending velocities along the radius, as shown in Fig. 4-1. Fig. 4-6 presents the burden descending velocity in the cross section at different levels. It can be seen that the velocities are almost uniform in the peripheral direc-

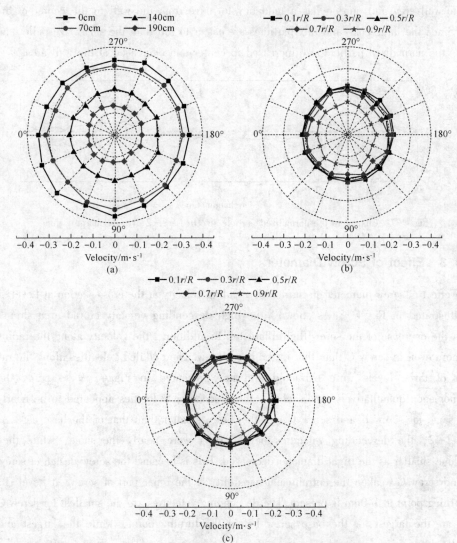

Fig. 4-6 Particle descending velocity along the peripheral directions
(a) Level-1; (b) Level-2; (c) Level-3

tions in Level-1, 2 and 3. In addition, the descending velocities at each level decrease along the radial direction and the velocity along the radius is relatively constant at higher level of SF, but it becomes uneven at lower level of SF. Fig. 4-7 shows the change of the normalized particle size, D/D_i, over the discharging time in Level-2, where the D is the average particle diameter in SF, and D_i is the initial average diameter. As can be seen, the normalized particle size remained nearly unchanged prior to $t = 3.4$s for three zones, then the normalized particle size became larger for the center zone, but decreased slightly in the middle and wall zones. This is because the velocities decrease from the center to wall area and burden has a tendency to move from the wall to the center of the SF. Since the large coke and flux particles are easier to roll than the small size pellet and ore, the normalized particle size increased in center and decreased at the wall area.

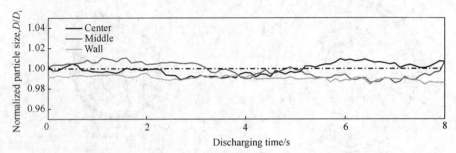

Fig. 4-7 Change in the normalized particle size in Level-2 over discharging time

4.1.3 Effect of Screw Diameter

The effect of screw diameter on burden descending velocity at the cross section at Level-1 is illustrated in Fig. 4-8. As shown, the high descending velocity of red area shrank with the decrease of the screw flight diameter. In addition, the velocity along the radius is more even in screw C than that in the base screw. The detailed velocities along the radius of three levels were extracted and are presented in Fig. 4-9 to study the phenomenon quantitatively. At Level-3, the descending velocities appeared to be nearly the same for screw B and screw C and they were smaller than that of the base case. At Level-2, the descending velocities beyond 0.7r were nearly the same, while they became smaller as the flight diameter decreased. This is because the screw design of screw B and screw C weaken the entrainment capability of the inner part of screw. At Level-1, a turning point at 140cm is present. The descending velocities are the smallest for screw C and are the largest for the base screw before the turning point, while the largest and smallest velocities appeared to reverse after the turning point. The velocity at 0cm in the case of screw C decreased by 23.9% when compared to the base case and the standard deviation decreased from 0.087 to 0.05. It can be concluded from these results that the velocity distribution along the radius became more even when the screw flight diameter was reduced.

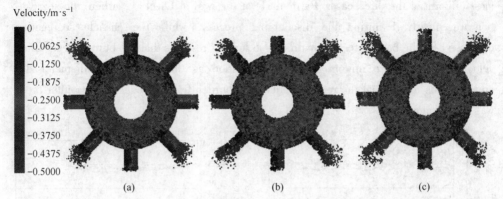

Fig. 4-8 Effect of screw diameter on burden descending velocity in the cross section at Level-1
(a) Base case; (b) Screw B; (c) Screw C

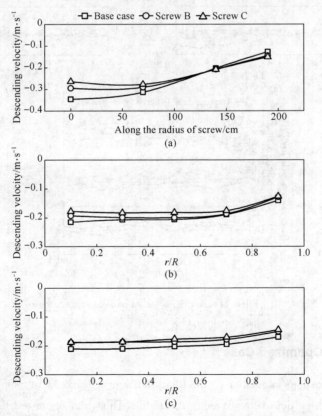

Fig. 4-9 Effect of screw diameter on particle velocity along the radius
(a) Level-1; (b) Level-2; (c) Level-3

Fig. 4-10 shows the effect of screw diameter on the particle segregation in SF. As can be seen, the particle segregation became larger in center and decreased in the wall area for various screws. However, the particle segregation in case of screw C was perhaps the most uniform of the three cases. Particularly at the wall at Level-1, where the segregation was nearly 1 during the discharging process, while the particle segregation decreased for the base case. The main reason for this result is that the burden descending velocity became more uniform in radial direction as the screw flight diameter decreased. This means that the rolling tendency of burden from the edge to the center areas was less pronounced.

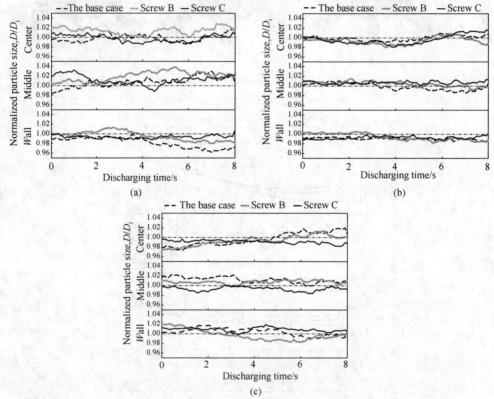

Fig. 4-10 Effect of screw diameter on particle segregation in SF
(a) Level-1; (b) Level-2; (c) Level-3

4.1.4 The Optimized Case

According to previous simulations, the flight diameter still has more room to decrease to achieve an even particle velocity along the radius. Thus, an optimized case was established by reducing the flight diameter and placing a guiding cone in the center of the bottom of the SF. The condition of optimized screw is tabulated in Table 4-3.

Table 4-3 Geometric properties of the optimized screw

Number	1	2	3	4	5	6	7
P/cm	37.5	39.4	43	46	51.8	62.9	73.3
D/cm	65	72.5	80	87.5	95	102.5	110

Fig. 4-11 shows the withdrawn volume and the net increment volume from No. 1 to No. 5 pitch for the optimized screw. As can be seen, the withdrawn volume increased along the screw and only 19.4% final screw transport volume was discharged by the first flight. In addition, the net increment volume also increased from the No. 2 pitch, which is a benefit for achieving a uniform pattern.

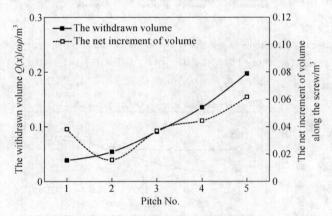

Fig. 4-11 Withdrawn volume and net increment volume calculated by Roberts analytic for optimized screw

Fig. 4-12 shows the solid flow patterns in the optimized screw. The initial positions of the traced layers appear to be the same as the base case. It can be seen that a relative uniform solid flow profile can be obtained in the optimized screw. The velocities along the radius between the base case and the optimized screw are compared in Fig. 4-13. The descending velocities along the radius for the optimized screw seem flatter than those of the base case at each level. For example, in Level-1, the difference between the maximum and minimum velocities in the optimized screw decreased by 71.6% compared with that of the base case. And the difference between the largest and the smallest descending velocities in Level-2 and Level-3 decreased from 0.075m/s to 0.014m/s and 0.04m/s to 0.019m/s, respectively. In addition, the standard deviations from Level-1 to Level-3 decreased from 0.087, 0.027 and 0.016 to 0.024, 0.018 and 0.015, respectively.

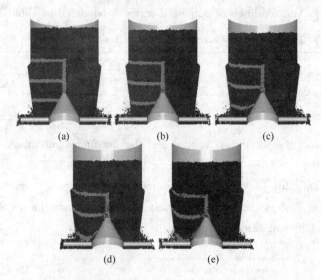

Fig. 4-12 Snapshots of cross-sections in the SF under the optimized screw
(a) $t=0s$; (b) $t=2s$; (c) $t=4s$; (d) $t=6s$; (e) $t=8s$

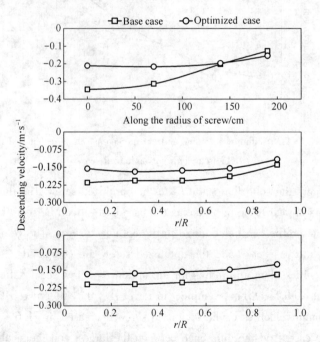

Fig. 4-13 Comparison of velocities between the base case and optimized screw along the radius

Fig. 4-14 shows a comparison of particle segregation between the base case and optimized screw at Level-1. It can be seen that the particle segregation in optimized screw tends to be more uniform than that in base case. The standard deviations in the center, middle and wall areas decreased from 0.0077, 0.0104 and 0.011 to 0.0067, 0.005 and 0.008, respectively. Therefore, the velocity distribution and particle segregation of the optimized screw appears to be more reasonable for even discharge of materials.

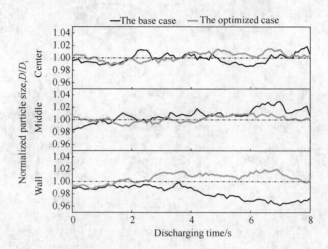

Fig. 4-14 Comparison of particle segregation between the base case and optimized screw at Level-1

4.2 Influence of Uneven Working of Screws

4.2.1 Simulation Conditions

The geometry used in this study is the same as shown in Fig. 4-1. The mass flow rate (namely, discharge rate) and the average descending velocity of particles were linear functions of the screw rotational speed from 200 to 1400 rpm (revolutions per minute) for horizontal screw conveyor. Thus, the rotational speed of screws was accelerated to 300 rpm to reduce the computational cost. For investigating the solid flow in a shaft furnace under uneven working of screws, the conditions of the screw feeders in shaft furnace were varied as shown in Fig. 4-15. The black screws denote the inactive screws. Cases A to C represent the adjacent non-working screws, and cases D to F represent the separated nonworking screws.

4.2.2 Effect of Adjacent Inactive Discharging

This section discusses the solid flow features at adjacent inactive discharging conditions. Fig. 4-16 shows the time-and stream-lines of solids on the vertical cross section at each inactive discharging condition. In base case of Fig. 4-16, the timelines show a

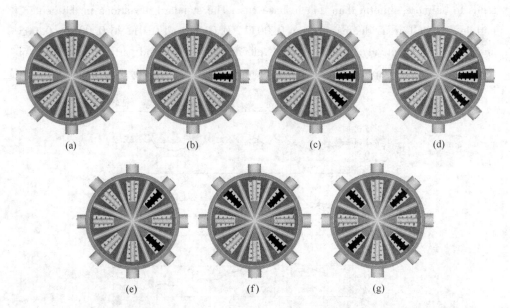

Fig. 4-15 Schematic diagram of geometriacl working condition of screws
(a) Base case; (b) Case A; (c) Case B; (d) Case C; (e) Case D; (f) Case E; (g) Case F

uniform descent in bustle zone, and form a 'U' shape below the bustle. As descending and discharging is continued, the timelines gradually change into a 'W' shape at the bottom of the shaft furnace. The streamlines in base case show that the particles charged nearly parallel to the wall descend linearly along the wall, and particles charged in the middle, move downward to the guiding cone before approaching the screws. However, for the imbalance conditions, the solid flow patterns are significantly altered, as shown in Fig. 4-16 by cases A, B and C. In all the conditions from case A to C, a sizeable stagnant zone can be seen on the right side, and the streamlines are not parallel to the wall and shift to the left. The timelines become curved in the bustle zone, and the descending velocity on the left side increases. The main difference among these cases is that the height of the stagnant zone on the right side increases from case A to C. The effect of inactive screws on the stagnant zone will be discussed in detail by considering the volume of the region where particles move extremely slowly and are even at rest.

Fig. 4-17 shows the percentage of stagnant zone volume in the cases with adjacent inactive discharging conditions. The percentage of stagnant zone volume is defined as the ratio of extremely slowly moving/motionless particles to all particles in shaft furnace. Clearly, the stagnant volume increases with an increasing number of inactive screws. However, when the number of inactive screws increases from one to three screws that are adjacent to each other, the stagnant volume in case C augments to 4.45 times

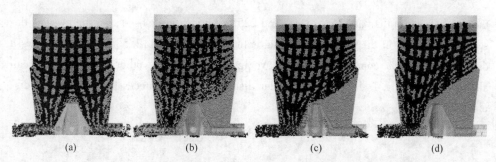

Fig. 4-16 Time-and stream-lines of solid on vertical cross section in
each inactive discharging condition
(a) Base case; (b) Case A; (c) Case B; (d) Case C

contrasting to case A. This increase is due to the three adjacent inactive screws decreasing the effect of the mechanical working of active screws, and the stagnant zone generated by each inactive screw links together.

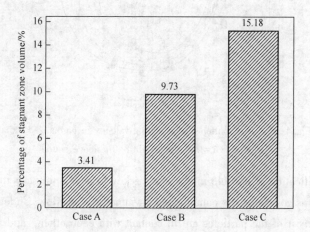

Fig. 4-17 Percentage of stagnant zone volume in the case with adjacent inactive screws

Fig. 4-18 shows the quantitative analyses of solid descending velocity in the normalized radial distance $r/R = 0.8$, at the height of 8.0m above the furnace bottom. It can be seen that the velocities are almost uniform in the peripheral direction for base case. However, for circumferential imbalance conditions, the descending velocities become quite non-uniform in the peripheral region. Because of the inactive discharging, two main features of the descending velocity are observed. The first observation is that burdens have relatively uniform descending velocities in the positions rotating clockwise from 90° to 270°, where the discharge process is normal operation. The velocities decrease with increasing the number of inactive screws. The other observation is that the lower descending velocities are seen above the inactive screws. For example, in case C, the discharging outlets at 135°, 180° and 225° positions are closed, so the average descending velocity in the positions rotating anticlockwise from 112.5° to 247.5° shows a decrease of 24.8% compared with the average velocity of the other positions. This can also be seen in Fig. 4-16 where a smaller distance between successive timelines is displayed by case C.

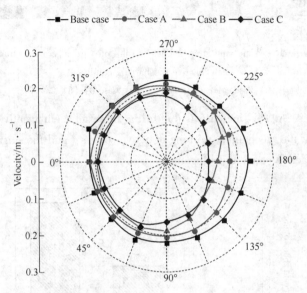

Fig. 4-18 Distribution of descending velocity in normalized radial distance $r/R=0.8$, at height 8.0m above bottom

Proper quantification of particle interactions is key to developing a better understanding of the mechanisms governing the complicated asymmetric solid flow. Under the current discharging rate, most of the particles are in contact with each other, and the contact force between particles can be spontaneously reconstructed. Fig. 4-19(a) shows the spatial distributions of interactive forces under different conditions. The interactive force is derived by the summation of forces at all contact points of a particle. It can be seen that particles exhibit a weak force network near the active screws for all cases. This is because in this zone, the particles are flowing fast and there are more voids and disconnections among the particles. In base case of Fig. 4-19(a), the largest interactive force exists at the top of the guiding cone. This is because particles need to support the burden above them. However, for the imbalance conditions, the largest interactive force can be observed in the stagnant zone where particles stay motionless or move very slowly. The area of largest interactive force increases from case A to C. The main reason for this trend is that with increasing the number of inactive screws, the size of stagnant zone increases, as shown in Fig. 4-16. Consequently, the number of particles with large contact forces increases. The probability density distribution of the interactive force for particles below the bustle zone under different conditions is shown in Fig. 4-19(b). The peak value decreases from 18.1% to 13.8% when three adjacent screws are out of action, which indicates that the number of weak particles decreases. In addition, the probability gradually diminishes with increasing interactive force, however, the probability density of larger force, such as 1500 ~ 2000N, increases with an increasing number of inactive screws, as shown in the bottom right region of Fig. 4-19(b). The average interactive force increases from 378N in base case to 519.5N in case C because a large stagnant zone has formed above the inactive outlets and more particles remain in direct contact with strong interaction force.

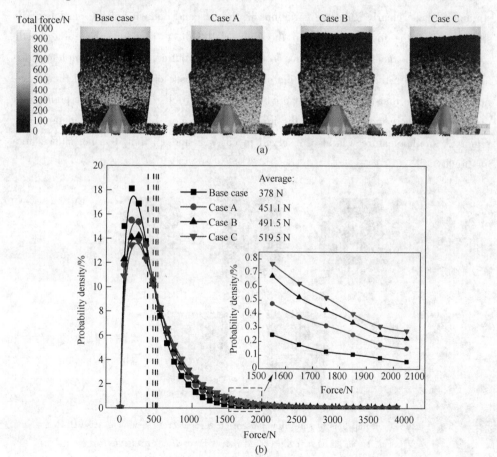

Fig. 4-19 Interactive force in shaft furnace under different conditions
(a) Spatial distribution; (b) probability density distribution

4.2.3 Effect of Separated Non-working Screws

To gain a comprehensive understanding of solid flow in a shaft furnace under circumferential imbalance conditions, the effect of separation of the non-working screws (such as cases D, E, and F shown in Fig. 4-15) on solid descending behavior is further discussed in this section. Fig. 4-20 shows snapshots of solid flow patterns with time-and stream-lines on the vertical cross section in case D. Although the active screws in Fig. 4-20 have the same rotating speed, the particles behave differently on each side of the center line. For the timelines, the particles on the left side descend faster than those on the right side. This is mainly caused by the two inactive screws being located around the active screw on the right side, which may reduce the particle downward motion. The streamlines also clearly illustrate variations in the descend behavior of the particles. The streamlines of the four tracers from the right wall shift to the left above the bustle zone. The fourth tracer can even move to the left screw outlet, which is completely different from that in the normal condition shown in base case of Fig. 4-16. All these phenomena indicate that the effect of inactive screws on particle velocity reflects changes in descending velocity not only above the inactive screw zone, but also at the neighboring region. A similar feature was also observed in cases E and F, but it is not shown here for brevity.

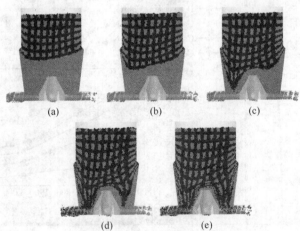

Fig. 4-20 Snapshots of solid flow pattern on vertical cross section in case D
(a) $t=60s$; (b) $t=69s$; (c) $t=78s$; (d) $t=87s$; (e) $t=96s$

Fig. 4-21 shows the solid descending velocity in the normalized radial distance $r/R = 0.8$, at the height of 8.0m above the furnace bottom under different separated inactive discharging conditions. The figure shows that the average descending velocity decreases with an increasing number of inactive screws, and as expected, the descending velocities above the inactive screws are smaller than those above the active screws. For example, in case E, three separated inactive screws are located in 135°, 225°, and 315°, and the lowest descending velocity can be seen at these positions. Moreover, the lowest descending velocity in case E is 0.158m/s, which is higher than that in case C which also has three inactive screws.

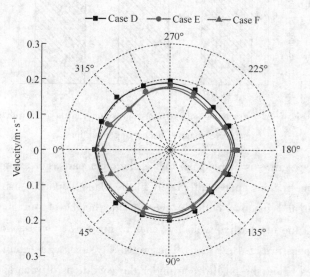

Fig. 4-21 Solid descending velocity in normalized radial distance $r/R=0.8$, at 8.0m height under different separate inactive discharging conditions

Fig. 4-22 shows the percentage of stagnant zone volume in the cases with separated inactive discharging conditions. Similar to that observed in the adjacent inactive conditions, the stagnant volume increases with an increasing number of inactive screws. However, the stagnant zone volume in the separated inactive conditions is lower than that in the adjacent inactive conditions when the number of inactive screws is the same. Further, it can be seen from Fig. 4-22 that the stagnant zone volume is almost proportional to the number of inactive screws. This is mainly because the inactive screws are not adjacent, so the stagnant zones are relatively independent, and the agglomeration effect of motionless particles doesn't gain sufficient development. This can also explain why the descending velocities in cases D and E are somewhat higher than those in cases B and C, respectively.

The distribution of the interactive force of particles on the horizontal cross section under different separated inactive discharging conditions is shown in Fig. 4-23(a), where the height is 4.0m above the furnace bottom. It can be seen that the interactive force

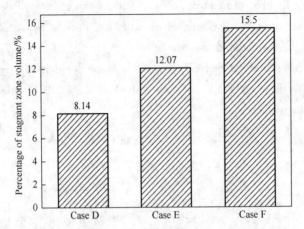

Fig. 4-22 Percentage of stagnant zone volume in the case with separate inactive discharging conditions

above the inactive screws is higher than that of the other parts. For example, in case F, high and low force regions appear alternately in the plane, and this can be clearly seen above the active and inactive outlets, even at the height of 4m above the furnace bottom. Notably, the low particle velocity zones in Fig. 4-21 correspond to the high interactive force zones in Fig. 4-23(a). The probability density distribution of interactive force for particles below the bustle zone under the conditions of separated non-working screws is shown in Fig. 4-23(b). It can be observed that the magnitude of interactive force varies largely for all cases. The average interactive force increases from 472.7N in case D to 538.6N in case F due to more particles remaining in direct contact with the strong interaction force in the stagnant zone. However, it should be pointed out that the average interactive force in the separated inactive discharging conditions, such as in cases D and E, is lower than that in cases B and C, respectively.

4.2.4 Effect of Discharge Rate

In this section, the effect of discharge rate on solid flow is investigated by increasing the rotation speed of active screws that are located near the inactive screws. For example, cases B and D are chosen as reference cases in this section, so the rotation speed of the two active screws near the inactive screw in case B and the three active screws near the two inactive screws in case D have an increase of 33% compared with that of the other active screws. The cases of high discharge rate for cases B and D are regarded as case B1 and case D1, respectively. Fig. 4-24 shows the effect of discharge rate on the percentage of stagnant zone volume. The figure indicates that with increasing discharge rate,

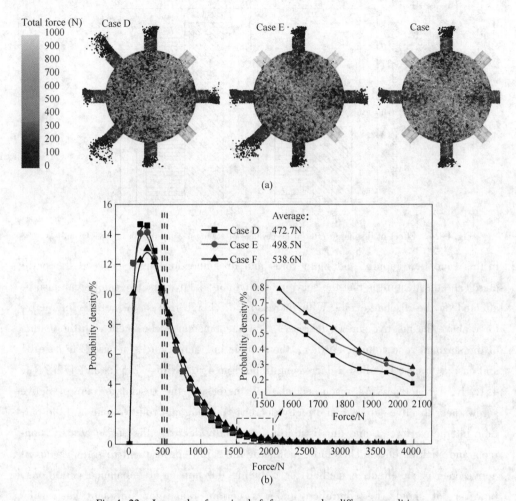

Fig. 4-23 Interactive force in shaft furnace under different conditions
(a) Spatial distribution on the horizontal cross section; (b) Probability density distribution

the stagnant zone volume in the shaft furnace decreases. The main reason for this trend is that at a high discharge rate, particle descending velocity is large, and the number of motionless particles is reduced. More particles become active near faster screws, which benefits the slow-moving particles located above the inactive screws to descend into the flow region.

Fig. 4-25 shows the effect of discharge rate on the distribution of solid interactive force on the horizontal cross section. Two main features can be observed as a result of the increase in discharge rate. The first feature is that the particle interactive force above the active screws decreases with increasing discharge rate. This is mainly because particles

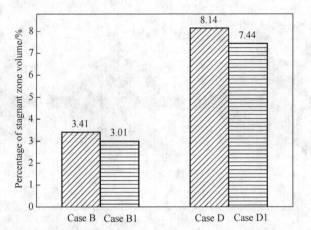

Fig. 4-24 Effect of discharge rate on percentage of stagnant zone volume in case B and D

in this zone flow rapidly and more voids and disconnections exist between the particles. The other feature is that the solid interactive force above inactive screws can also be affected by the discharge rate. With increasing discharge rate, the particle interactive force above the inactive screws decreases. The phenomenon is consistent with the feature of the stagnant zone volume. Further, the average interactive forces of case B to case B1 and case D to case D1 show a decreasing trend from 451.1N to 442.87N and 472.7N to 451.4N, respectively. As discussed above, increasing the discharge rate of active screws near inactive screws can directly decrease the stagnant zone volume and the particle interactive force, and thus is able to potentially decrease the period of static contacts and sticking effect. Therefore, adjusting the screw speed of active screws near inactive ones is an effective method for reducing the impact of imbalance conditions, however, it should be controlled under a certain range for sufficient gas – solid reduction reactions.

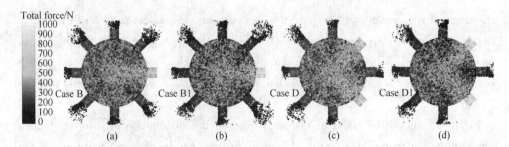

Fig. 4-25 Effect of discharge rate on distribution of interactive force on horizontal cross section
(a) Case B; (b) Case B1; (c) Case D; (d) Case D1

4.3 Summary

A three-dimensional actual size model for simulating the influence of screw on burden descending velocity and particle segregation in COREX shaft furnace was developed using DEM. The conclusions can be summarized as follows:

For the influence of screw design, in the base case, burdens are drawn primarily down from the first flight of the screw and the largest descending velocities are observed above the tip of the screw, while the sluggish moving particles are located above the end of the screw. Burden descending velocities are almost uniform in the peripheral direction and decrease along the radial direction in the base case. Besides, the normalized particle size increased in center area and decreased in wall area. Reducing the flight diameter of the screw restrains the descending velocities in the central area of the SF and helps to obtain an even flow pattern. The rolling tendency of the burden from edge to center areas also appeared to decrease with a decrease in the flight diameter. A relative uniform solid flow profile can be obtained in the optimized screw and the evenness of descending velocity along the radius has been improved greatly.

For the influence of uneven working of screws, under abnormal conditions where several inactive screws are adjacent to each other, the solid flow patterns change significantly. A large stagnant zone forms above the inactive screws, and the stagnant zone volume increases with an increasing number of inactive screws. Microscopic analysis confirms the particle interactive force increases with an increasing number of inactive screws. Under asymmetric conditions, where inactive screws are separated from each other, the descending velocities above inactive screws are smaller than those above active screws. However, the average descending velocity, stagnant zone volume, and interactive force in separated inactive conditions are lower than those in adjacent inactive conditions when the number of inactive screws is the same. Increasing the rotation speed of active screws that are located near the inactive screws can decrease the stagnant zone volume and particle interactive force in shaft furnace. Therefore, adjusting the screw speed of active screws near inactive screws is an effective method for reducing the imbalance impact, however, it should be controlled under a certain range for sufficient gas-solid reduction reactions.

References

[1] Owen P J, Cleary P W. Prediction of screw conveyor performance using the Discrete Element Method (DEM) [J]. Powder Technology, 2009, 193: 274~288.

[2] Hou Q F, Dong K J, Yu A B. DEM study of the flow of cohesive particles in a screw feeder

[J]. Powder Technology, 2014, 256: 529~539.
[3] Shimizu Y, Cundall P A. Three-dimensional DEM simulations of bulk handling by screw conveyors [J]. Journal of engineering mechanics, 2021, 127: 864~872.
[4] Fernandez J W, Cleary P W, McBride W. Effect of screw design on hopper drawdown of spherical particles in a horizontal screw feeder [J]. Chemical Engineering Science, 2011, 66: 5585~5601.

5 Gas-solid Flow in a Large-scale COREX Shaft Furnace with Center Gas Supply Device Through CFD-DEM Model

Over the past decades, micro-scopic and macro-scopic properties of gas-solid flow in blast furnace has been extensively investigated by combining discrete element method (DEM) with computational fluid dynamics (CFD). The successful application of CFD-DEM prompts a detailed simulation and analysis of gas-solid flow in COREX process by this approach. For example, the effects of packing way and particle diameter on the gas flow in a shaft furnace were studied using a hybrid model of CFD-DEM coupling. Hou et al. studied the effects of discharge rate and two common burden structures on the thermal behavior of gas-solid flow in the shaft furnace. However, these works only considered single size or two kinds of particles. As a result, the particle size segregation behavior under different burden profiles, which is the principal factor affecting void distribution and gas flow, cannot be studied.

In this chapter, the research object is the latest COREX shaft furnace with central gas distribution (CGD). The influences of CGD on gas-flow in COREX shaft furnace were investigated through CFD-DEM model. The comparison of particle velocity and segregation in the shaft furnace with and without CGD was investigated. Then the voidage distribution and gas flow in shaft furnace was further studied. The effects of CGD on the average residence time, variance, and volume fraction of the gas and solid phases were also discussed. Furthermore, the effect of burden profile on gas-solid distribution in SF with CGD were discussed.

5.1 CFD-DEM Model

Cundall and Strack originally proposed DEM for particles, and particle motion can be solved by the Newton's second law. The governing equations of particle i with mass m_i, and rotational inertia I_i can be written as

$$m_i(\mathrm{d}\boldsymbol{v}_i)/\mathrm{d}t = \sum_{j=1}^{k_i} (\boldsymbol{F}_{\mathrm{cn},\,ij} + \boldsymbol{F}_{\mathrm{dn},\,ij} + \boldsymbol{F}_{\mathrm{ct},\,ij} + \boldsymbol{F}_{\mathrm{dt},\,ij}) + \boldsymbol{F}_{\mathrm{pf},\,i} + m_i \boldsymbol{g} \quad (5-1)$$

$$I_i(\mathrm{d}\boldsymbol{\omega}_i)/\mathrm{d}t = \sum_{j=1}^{k_i} (\boldsymbol{T}_{ij} + \boldsymbol{M}_{ij}) \quad (5-2)$$

where, v_i, ω_i and k_i are the translational velocity, rotational velocity and the particle numbers which are contacted with particle i, respectively. $m_i g$, $F_{pf,i}$, $F_{cn,ij}$, $F_{dn,ij}$, $F_{ct,ij}$, and $F_{dt,ij}$ respectively represent the gravitational force, the particle-fluid interaction force, the normal contact forces, the normal damping forces, the tangential contact force and the tangential damping forces. In addition, the torque acting on particle i includes two components: tangential torque $M_{t,ij}$ and rolling friction torque $M_{r,ij}$, generated from the tangential force and rolling friction, respectively.

The continuous fluid flow was calculated from the continuity and Navier-Stokes equations based on the local mean variables over a computational cell level, given by

$$\partial(\varepsilon\rho)/\partial t + \nabla \cdot (\rho\varepsilon u) = 0 \tag{5-3}$$

$$\partial(\varepsilon\rho u)/\partial t + \nabla \cdot (\rho\varepsilon uu) = -\nabla p + \nabla \cdot (\mu\varepsilon \nabla u) + \rho\varepsilon g - F_{pf} \tag{5-4}$$

where ε, ρ, p, u, and μ respectively are the void fraction, density, pressure, velocity vector and viscosity of gas phase. $F_{pf}\left(=\dfrac{\sum_{i=1}^{n}F_{pf,i}}{V_{cell}}\right)$ is the volumetric particle-fluid interaction force. The porosity was calculated through $\varepsilon_i = 1 - \sum_{i=1}^{k_v}\dfrac{V_i}{\Delta V}$, where V_i is the particle volume in a mesh, ΔV is the volume of the mesh.

The drag model used in this work is similar to the previous work for studying the gas-solid flow in reduction shaft of COREX. The average Reynolds number in COREX shaft furnace is about 3000, which is suitable for the C_d in this work. The coupling of CFD-DEM methods could be very different in different situations. The coupling between the particle and gas phases appears through the particle volume fraction and the inter-phase force in the gas-phase momentum equation which represents momentum exchange between the particle phase and the gas phase. At each time step, DEM will give information, such as the positions and velocities of individual particles, for the evaluation of porosity and volumetric gas drag force in a computational cell. CFD will then use these data to determine the gas flow field which then yields the gas drag forces acting on individual particles. Incorporation of the resulting forces into DEM will produce information about the motion of individual particles for the next time step. In this work, the update interval of the gas flow distribution was set to once every 10 steps of the DEM model, based on the convergence property and the accuracy of the calculation. The time step of DEM is 5.0×10^{-5} s and the time step of CFD is 5.0×10^{-4} s.

5.2 Model Validity

Firstly, the model was validated by comparing the angle of repose of non-spherical coal particles. A multi-sphere model is used to construct these particles. There are four typical morphologies of the coal particles, named Sa (Shape A) to Sd (Shape D) as shown in Fig. 5-1. The mass fractions of Sa to Sd in experiment and simulation are the same, which are 17.3%, 16.6%, 59.6% and 6.5% respectively. Fig. 5-1 presents the angle of repose and mean voidage in experiment and simulation. The experiment results show that the angle of repose is 29°, and the mean voidage of the pile is 0.472. It seems the appearance of the experimental and simulated coal piles does not look very similar, this is mainly due to the different shooting angle. Actually, in the simulation, the angle of repose and mean voidage of coal are 29.1° and 0.454 respectively. The simulated coal pile behaviors are basically consistent with the experimental data.

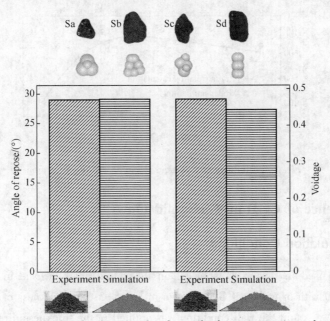

Fig. 5-1 Comparison of angle of repose of coal particles between experimental and simulation

In order to further verify the model, this work compares the pressure drop of non-spherical particle packed bed in theoretical calculation and simulation. The Ergun equation $\dfrac{\Delta P}{L} = \dfrac{150\mu v (1-\varepsilon)^2}{[(d_e \cdot \phi)^2 \cdot \varepsilon^3]} + 1.75 \dfrac{\rho_g v^2 (1-\varepsilon)}{[(d_e \cdot \phi) \cdot \varepsilon^3]}$ was used to calibrate the model.

The particles with different non-convexity shape factors but having the same equivalent diameter are shown in Fig. 5-2(a). The packed bed composed of the same number of non-spherical particles is shown in Fig. 5-2(b). Fig. 5-2(c) presents the variation of pressure drop in packed bed with the particle shape. It can be seen when the shape factor is less than 0.25, the pressure drop increases with the increase of the shape factor. Then the pressure drop decreases with further increase in the shape factor. The main reason is that when the shape factor is greater than 0.25, the non-spherical particles are easily to form a looser packing structures, thereby reducing the pressure drop. The theoretical calculated pressure drops from the Ergun equation agree well with the simulated results.

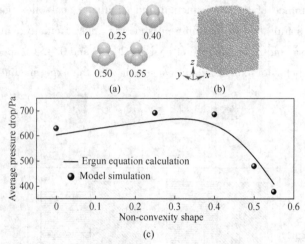

Fig. 5-2 Influence of particle shape on the pressure drop

5.3 Influence of CGD on Gas-solid Flow

5.3.1 Simulation Conditions

The geometry used in this work is based on the actual shaft furnace in a commercial COREX-3000 with or without CGD, shown in Fig. 5-3. The thickness of the slot model equal to $6.5d_{coke}$, and the wall condition was applied in the thickness direction. The geometry is divided into hexahedral mesh, and the diagonal of the smallest grid equal to $2d_{pellet}$. In this work, the discharge rate is 610 particles per second. And the inverse 'V' shape profile is form to investigate the gas-solid flow features in COREX shaft furnace with and without CGD. The simulation starts with the random generation of a certain number of well-mixed particles to form a packing bed in this model shaft furnace. During charging process, the burden profiles is formed and the particle segregation can be observed. Then the burden is discharged from the bottom at the pre-set rate. When the top surface of burden reaches stock line level, burden is charged alternately. After the burden descending behavior is steady, gas is injected from the slot or CGD. In this work, the inlet velocity in shaft without CGD is consistent with the previous study of 20m/s through the slot gas inlet, while the slot gas inlet velocity is 12m/s and the CGD gas inlet velocity is 8m/s in the furnace with CGD by calculation. The outlet boundary of gas phase is Outflow.

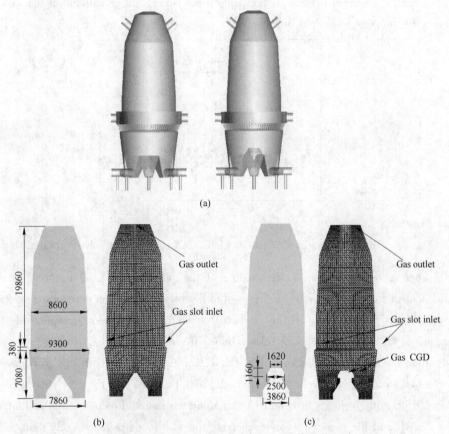

Fig. 5-3 The geometry used in this work (Unit. mm)
(a) Schematic diagram of COREX shaft furnace; (b) Slot model and computational grid for the shaft without CGD; (c) Slot model and computational grid for the shaft with CGD

5.3.2 Particle Velocity and Segregation

In this section, 5 samples are collected along the radius direction at different height for the exploration of particle descending and segregation behavior. For particle velocity, the particle average velocity in each sample can be obtained. And the total mass of each burden in each sample is collected in order to calculate the particles segregation. The detailed analyses are as follows.

Fig. 5-4 shows the variation of velocity of particles with time in COREX shaft furnace with or without CGD while the particles reach a steady state. The variation tendency of both two furnace structure is similar, that is, the velocity of particles reaches the sharp peak value with the burden charging, which may have an impact on the moving particles, and thus cause the fluctuation of velocity; the narrow fluctuation range of velocity is observed in the interval between the peaks; moreover, the magnitude of the average velocity for two cases is slight difference.

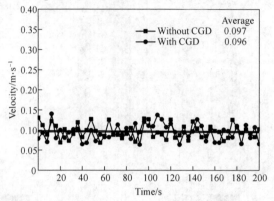

Fig. 5-4　The velocity variation of particles in COREX shaft furnace without or with CGD

In order to have a further understanding of the influence of CGD on solid flow in COREX shaft furnace, three levels are used to represent the burden descending velocity, at the height of 15.0m, 10.0m and 6.0m above the bottom, respectively. The comparison of burden descending velocity along the radius direction between cases with and without CGD is shown in Fig. 5-5. The velocities investigated in this work are the average velocities of relatively stable period. From Fig. 5-5 (a), it is clear that the burden descending velocity at height 15.0m along the radius direction is quite similar in both cases, and the particles velocity close to the wall is slower apparently than that of in furnace center due to the boundary effect. This phenomenon indicates that the installation of CGD in the shaft furnace have little influence on burden descending velocity in the upper part of shaft furnace.

The burden descending velocity at height 10.0m is shown in Fig. 5-5(b). It can be seen that, the velocities of two cases reveal the difference. The velocity evolution of the case without CGD is quite similar with the results of the height 15.0m. However, it should pay attention to the case with CGD, due to the inhibition of CGD, the particles velocity in the center is quite smaller than that of the case without CGD, while the particles near the wall fall down faster. Although the velocity shows difference along the radius direction, the average velocities are similar. What's more, the descending velocity of particles located in the upper part of the screw is the highest in both two cases.

Fig. 5-5(c) shows the burden descending velocity at height 6.0m, which is close to the top of CGD. The difference of velocities evolution between the cases with or without CGD is apparent. With the effect of guiding cone, the falling of particles in the furnace center was blocked, and thus the velocity is quite slow in both two cases; and the velocity in case with CGD is slower obviously because of the further influence of CGD. It's worth noting that in the case with CGD, the descending velocity close to the wall is the fastest along the radius direction, that may be due to the combined action of the burden profile and the mounting of the CGD.

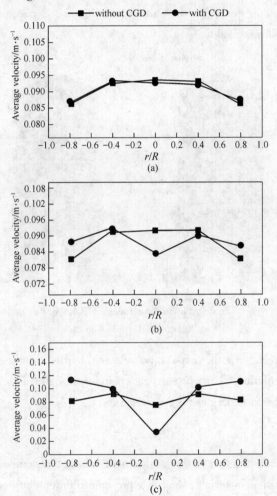

Fig. 5-5 Burden descending velocity along the radius direction at different height above bottom in shaft furnace without or with CGD
(a) 15.0m; (b) 10.0m; (c) 6.0m

Then the particle distribution in shaft furnace with or without CGD is explored and displayed in Fig. 5-6. As the burden distribution changes slightly at stable condition, the instantaneous state selected randomly is used in this work. Different burdens filling in the shaft furnace are labeled with different colors, where pellet, ore, coke and flux particles are gray, blue, green and red color, respectively. As shown in Fig. 5-6, inverse 'V' shape burden profile is generated in the upper of the furnace in both cases with or without CGD, and the particle segregation is obvious. The large particles such as coke and ore are on the edge of the furnace, while the center of shaft furnace is mainly filled with pellet. Moreover, this phenomenon is some more prominent in the case with CGD. In order to further quantitate the particle distribution, a particle segregation index is defined in the following study.

Fig. 5-6 The distribution of particles in COREX shaft furnace
(a) Without CGD; (b) With CGD

For particle segregation, the mass ratio of each burden in the above-mentioned sample can be calculation. Then, a segregation index is defined to assess the particle segregation, and the formula can be descripted as,

$$N_k = \frac{P_k - P_{0,k}}{P_{0,k}} \quad (5-5)$$

Where, P_k is the mass fraction of the burden materials in the sample, $P_{0,k}$ is the initial mass fraction of burden materials, and k represents different burden. It's obvious that the N_k reflects the segregation degree, that is, the larger absolute value of N_k would bring out a more obvious particle segregation phenomenon.

Fig. 5-7 shows the comparison of particle segregation along the radius direction between two cases at the height of 15.0m and 6.0m above the bottom. As shown in Fig. 5-7(a),

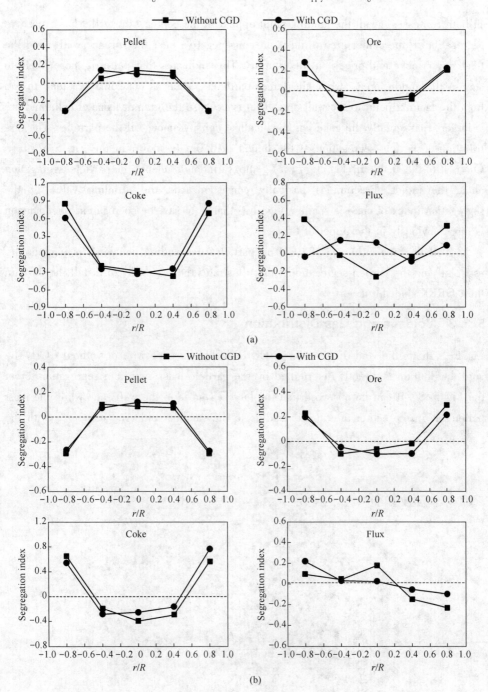

Fig. 5-7 Particle segregation along the radius direction at different height above bottom in shaft furnace with or without CGD

(a) 15.0m; (b) 6.0m

in both two cases, with the landing point appear in the center, the pellet has a positive segregation in the center area while it presents negative near the furnace wall, and the tendency of coke and ore is on the contrary. This indicates that pellet is more likely to stay in the middle area, while the larger particles, such as coke and ore tend to roll from the landing point to the wall area. That is because the granular size of coke and ore is larger, particularly the coke has the smallest density among all the particles. The distribution of particle segregation at the height of 6.0m is shown in the Fig. 5-7(b). Contrast to the result in Fig. 5-7(a), the variation tendency of particle segregation along the radius direction is basically consistent. And the absolute value of the segregation index of coke is smaller slightly, that is to say the segregation phenomenon is more obviously in the upper of furnace.

As a whole, the influence of CGD on particles descending behaviors is inapparent, especially on the burden segregation, that illustrates the feasibility to install the CGD in the COREX shaft furnace.

5.3.3 Voidage and Gas Distribution

Fig. 5-8 shows the void distribution in COREX shaft furnace with or without CGD. The void distribution is directly determined by the particle mass ratio and segregation. From the Fig. 5-8, it can be observed that the largest void is in the wall zone while the most pellet stay in the center area, and leading to the smaller void. And compared with the

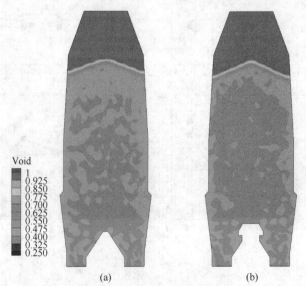

Fig. 5-8 Void distribution in COREX shaft furnace
(a) Without CGD; (b) With CGD

furnace without CGD, the overall voidage of shaft furnace with CGD is somewhat smaller, particularly in the middle zone.

To further study the effect of CGD on the gas flow, the gas was injected from the slot or CGD of the shaft furnace after the burden was steady. The streamline of the gas flow with or without CGD is shown in Fig. 5-9. Compare Fig. 5-9(a) with (b), the gas distribution injected from the slot of the furnace show a 'J' shape. And for the case with CGD, the gas was injected from both slot and CGD of the furnace, and that enriches the central gas distribution of the lower part of the shaft furnace.

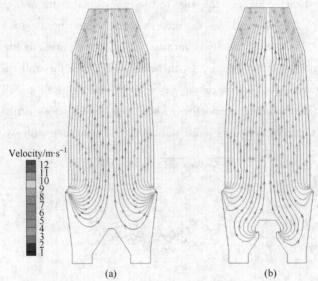

Fig. 5-9 The streamline of the gas flow in COREX shaft furnace
(a) Without CGD; (b) With CGD

The gas radial distribution at different height is discussed for the further investigation of gas flow in the shaft furnace. And Fig. 5-10(a) shows the radial distribution of gas velocity at height 5.5m that in the upside of the CGD. The gas distribution of two cases vary dramatically due to the influence of the installation of CGD. For the case with CGD, the gas velocity is increased initially and then decreased from center to wall, while the gas distribution of furnace without CGD vary significantly. It can be observed that the gas velocity of furnace with CGD in the middle zone (the absolute value from the center is in the range of 1m to 2.7m) is much larger than that of the case without CGD, while in the wall area, the case without CGD has the larger velocity. All that indicated that the influence of the CGD on the gas distribution below the slot of furnace is significant, it can enhance the gas flow in the center of the furnace and thus may improve the reduction rate of furnace.

Fig. 5-10(b) reveals the gas radial distribution at height 8m (near the upper of the slot). For the condition with CGD, the velocity increases with the increase of height, for example, the gas velocity increases from 3.44m/s to 6.04m/s in the center area. The variation tendency of two cases is mainly consistent, that is, the velocity near the wall area is larger than that of in the center zone, but the interval between maximum and minimum of furnace with CGD is slightly narrower. The main reason is the 'wall effect'. As the void fraction near the walls are larger than that in the center zone, the gas velocity is higher near the walls. Moreover, the gas velocity in the middle area of that case is significantly higher while that of two cases is adjacent near the wall. As seen in Fig. 5-10(c), the overall gas velocity increases when gas continues to move upwards. At height 15m, the velocity of furnace with CGD is obviously higher than that of the case without CGD. What's more, by the contrast between the wall and center gas velocity for both two cases, the maximum gas velocity near the wall are 19% and 12.6% higher than that in the center respectively, that is to say the gas distribution is more uniform and the central gas flow is more sufficient in the furnace with CGD.

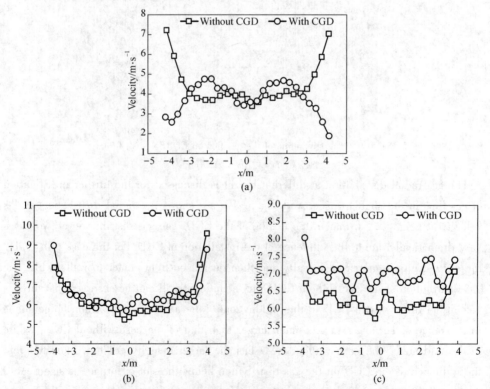

Fig. 5-10 The radial distribution of gas velocity at different height
(a) 5.5m; (b) 8m; (c) 15m

As a whole, the overall gas velocity is improved and the center gas distribution is promoted apparently with the installation of CGD, and thus the metallization rate and gas utilization ratio may further be improved. The pressure drop in the shaft furnace with or without CGD is further explored to comprehend the effect of CGD on gas flow broadly. The isobar distribution of both two cases is shown in Fig. 5-11. It's obvious that the isobar is basically horizontal for both two cases in the upper and middle part of the furnace. And it presents 'U' shape near the slot inlet, the largest pressure drop is at the slot zone in the condition without CGD, while that is at the lower part of the furnace with CGD. What's more, it can be seen that the furnace with CGD has the larger pressure drop, and its largest pressure drop is 6696Pa while that of the case without CGD is just 5939Pa. The longitudinal pressure drop at different position from center in shaft furnace is described in Fig. 5-12. Among them, Fig. 5-12(a) and (c) are the results on the left and right side of the symmetric position in the furnace, and Fig. 5-12(b) depicts the longitudinal pressure drop at center position. On the one hand, it can be seen that the pressure drop in shaft furnace is extraordinary symmetric, which is consistent with the result of Fig. 5-11. On the other hand, when the height exceeds 6m (above slot of the furnace), the pressure drop increases linearly along the longitudinal direction for all the position, and the pressure drop reaches the maximum at the slot level for the position of 3.8m from the center in the furnace without CGD, while the pressure drop continues to rise gently in another case. Furthermore, the pressure drop in the

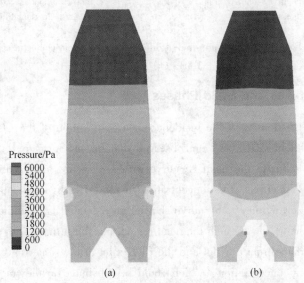

Fig. 5-11 The pressure drop in COREX shaft furnace
(a) Without CGD; (b) With CGD

furnace with CGD is higher than that of the case without CGD consistently for all the position, this is because the gas injected from the CGD undergoes the higher packing bed when moving upward.

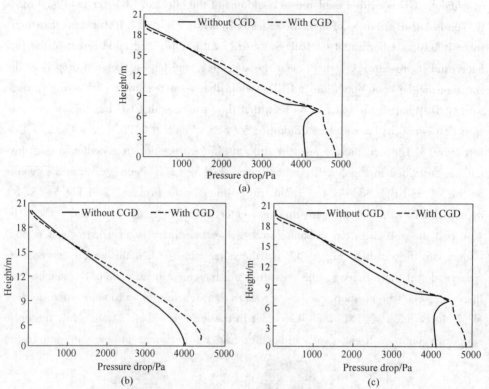

Fig. 5-12 Longitudinal pressure drop at different position from center in shaft furnace
(a) -3.8m; (b) 0m; (c) 3.8m

5.3.4 RTD of Gas and Solid Phases

Fig. 5-13 shows the density of the residence time distribution of the COREX shaft furnace with and without CGD. One can observed that for the furnace without CGD, the RTD curve was similar to the ideal C-curve with a plug flow and a mixed flow in series. In such an RTD curve, the concentration increased vertically, followed by an exponential decay. Furthermore, the curve showed an extended tail, indicating the existence of a slow moving flow through the dead region. For the furnace with CGD, the tracer entered the furnace from the slot and the CGD inlet at the same time. There were two peaks of the tracer concentration, which could be attributed to the tracer from the gas slot inlet and the CGD inlet reaching the monitoring surface successively.

The influence of CGD on the average residence time and variance in the COREX shaft

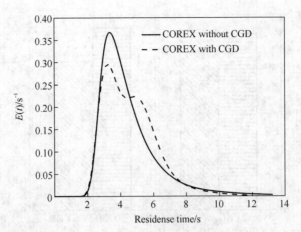

Fig. 5-13 Density of residence time distribution of COREX with and without CGD

furnace is shown in Fig. 5-14. Upon the addition of CGD, the average residence time increased from 4.48s to 4.6s. It is expected the CGD inlet gas to travel longer in the furnace, resulting in an increase in the overall residence time. The variance of the residence time showed that the value decreased from 3 to 2.44s^{-2} upon the addition of CGD. The dimensionless variance, which referred to the variance divided by the square of the average residence time, was reduced from 0.15 to 0.115. The variance or dimensionless variance is often used to evaluate the degree of deviation of random variables from the mean value, which corresponds to the flow pattern one to one. In the case of a plug flow, the variance is the smallest, while in the case of a full mixed flow, the variance is the largest. Therefore, the gas flow in the COREX shaft furnace was a non-ideal flow. Upon the addition of CGD, the dimensionless variance decreased and the deviation from the plug flow became smaller. One of the main reasons for this was that the CGD could realize the central gas supply and make the gas distribution more uniform between the wall and the center of the shaft furnace, which was conducive to the formation of a plug flow. In contrast, the initial gas velocity and the momentum in the gas inlet in the shaft furnace with CGD decreased, which made the mixing degree in the shaft smaller and the flow pattern closer to the plug flow.

Fig. 5-15 presents the calculated participations of flows in the shaft furnace with and without CGD. It can be seen that upon the addition of CGD, the dispersed plug flow increased from 58.7% to 60.9%, increasing by 3.7%. This result was also verified in a previous work, but the increase in the volume fraction of the plug flow region was smaller. In the gas-solid counter-current reactor, the larger the dispersed plug volume fraction was, the better was the gas-solid transport and reaction, thus promoting the effi-

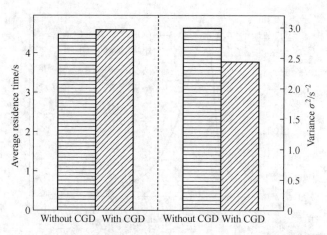

Fig. 5-14 Effect of CGD on the gas average residence time and variance

cient and stable operation of the reactor. In real practice, the metallization rate of the DRI of COREX shaft furnace with CGD improved. As the reducing gas was injected through the CGD inlet, the dead volume fraction at the bottom of the furnace could be reduced effectively. Hence, the dead volume fraction in the shaft furnace with CGD was low, only 2.9%. With the installation of CGD, the well-mixed volume fraction increased from 34.13% to 36.2%.

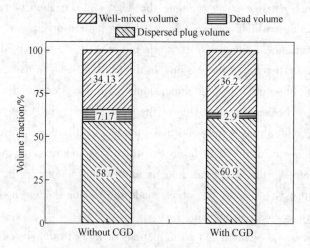

Fig. 5-15 Calculated participations of the flows for the COREX shaft furnace with and without CGD

The solid phase RTD curves for the shaft furnace with and without CGD are shown in Fig. 5-16. Both the RTD curves presented a typical C-type. It can be seen that the solid residence time ranged from 171.5s to 267.5s for the case without CGD, and the largest probability of the residence time appeared in 177.5s. However, in the shaft furnace with CGD, the RTD curve shifted to the left. The residence time ranged from 160s to 263s, and the peak of the curve reduced to 168s. Both the curves diminished gradually with an increase in the time, which was mainly attributed to the solid sluggishly descending near the vessel wall. These slow-moving particles showed the potential to increase the static contact, and a scaffold might be formed. Note that the average solid residence time in the numerical model was considerably smaller than that in real practice. However, the macroscopic descending behavior in the model was consistent with that in real practice. In addition, at the same discharge rate, the change in the solid residence properties of the shaft furnace with and without CGD could be compared and the effect of CGD could be further evaluated.

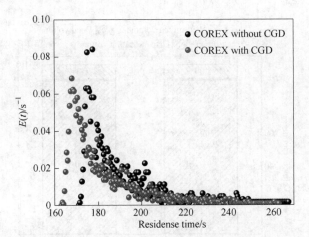

Fig. 5-16 Density of residence time distribution of solid in COREX with and without CGD

Fig. 5-17 shows the solid average residence time and variance in COREX shaft furnace with and without CGD. The average residence time of the solid phase decreased from 191.3s to 183.1s when the shaft furnace was designed with the CGD. The main reason for this was the compression of the effective furnace volume by the installation of CGD; further, the solid descending velocity was higher than that of the furnace without CGD. The variance showed a larger value in the case with CGD, which indicated that the deviation from the plug flow increased. This was mainly attributed to the fact that the CGD at the bottom center could increase the solid back-mixing intensity.

A comparison of the solid flow characteristics in the shaft furnace with and without the CGD is shown in Fig. 5-18. It can be observed that the dispersed plug volume frac-

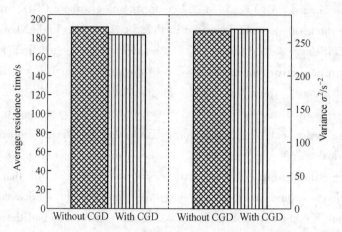

Fig. 5-17 Effect of CGD on the solid average residence time and variance

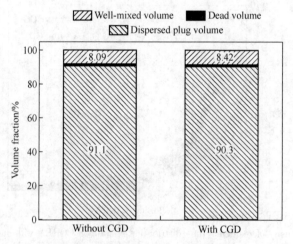

Fig. 5-18 Comparison of the calculated participations of the solid flows in shaft furnace with and without CGD

tion decreased for the furnace with the CGD. This was the expected effect of the modifications of the bottom structure of the shaft furnace. With the introduction of the CGD, the cross-sectional area at the bottom of the shaft furnace varied significantly and the solid flow was more complicated. As the area under the RTD curves from the time, $\theta=2$ to ∞ was close to zero, the dead volume analyzed by the combined model of CCMT and CPFR was zero. However, some tracer particles stagnated above the bottom center of the shaft center and did not flow out the reactor. Thus, in this work, the dead volume fraction was defined as the ratio of the extremely slowly moving/motionless particles to all particles in the shaft furnace. With the introduction of CGD, the dead volume fraction increased from 0.81% to 1.28%. The retardation of the flow area due to the CGD installed at the bottom center increased the number of motionless particles at the upstream of the CGD. The well-mixed volume fraction in the shaft furnace with CGD was 8.42%, which was 4% higher than that in the shaft furnace without CGD.

5.4 Influence of Burden Profile on Gas-solid Flow

5.4.1 Simulation Conditions

In the actual charging process, multi-scale burdens will segregate under different charging matrices, which will affect the burden profile and the structure of packed bed, and then impact the gas distribution in the furnace. Hence, five burden profiles, including 'V' shape, inverse 'V' shape, 'W' shape, 'M' shape and 'Flap' shape are studied in this work, as shown in Fig. 5-19. Firstly, a certain number of uniformly mixed particles are randomly generated in the model shaft furnace to form a packed bed. For different burden profile, in this work, it is mainly realized by changing the charging point. For example, the 'V' shape burden profile is obtained by charging the mixed particles to the point near the furnace wall through a chute, so that the burden spontaneously forms a pile with the tip near the furnace wall and the pile foot at the center of the furnace. The burden is then discharged from the bottom at a pre-set rate, the discharge rate is 610 particles per second. When the top surface of the packed bed reaches stock line level, burden is charged alternately. After the burden descending runs

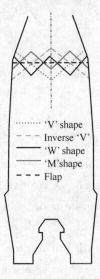

Fig. 5-19 Schematic of burden profiles

stably, the gas is injected both from the slot inlet and CGD inlet. In the present calculation, the slot gas inlet velocity is 12m/s and the CGD gas inlet velocity is 8m/s. The outlet boundary of gas phase is Outflow.

5.4.2 Burden Descending Velocity and Particle Segregation

Fig. 5-20 shows the transient burden descending velocity at the same time in COREX shaft furnace with CGD under different burden profiles. Although the transient velocities under different burden profiles are not completely consistent at the same time, there are two main characteristics can be observed in all cases. One is that the rapidly descending zone is located above the discharge outlets and extends to the top edge of the CGD. This zone can alleviate the issue of adhesion and agglomeration caused by the reduction of iron burdens promoted by the injection of reducing gas from CGD inlet. The other feature is that the minimum descending velocity exist at the top of CGD.

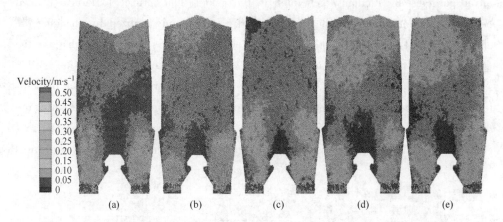

Fig. 5-20　Transient burden descending velocity in COREX shaft furnace with CGD
(a) 'V' shape; (b) Inverse 'V' shape; (c) 'W' shape; (d) 'M' shape; (e) 'Flap' shape

In order to have an overall understanding of the influence of burden profile on the solid movements in shaft furnace, the average velocity over a period of time (2s) at two height levels is used to characterize the descending velocity. Fig. 5-21 shows the quantitative analyses of the solid descending velocity. The evolution of the solid descending velocity at each height in the furnace under different bed shape are basically similar. At $h=15$m, as shown in Fig. 5-21(a), all cases present that the descending velocity in the center is slightly higher than that in the wall zone, and the overall average velocities are around 0.09m/s. With the downward movement of solid, the descending velocity is mainly affected by the screw feeders. From Fig. 5-21(b), it can be seen the velocities of the cases with different burden profiles are quite similar, showing a low velocity in center zone and a high velocity in the wall zone. All these phenomena indicate that the burden profile has a minor effect on the solid descending velocity in shaft furnace.

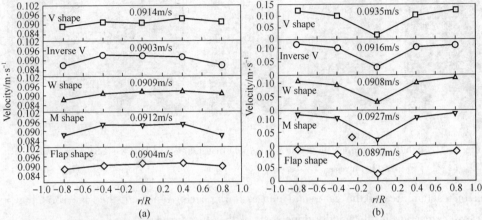

Fig. 5-21 Influence of burden profiles on descending velocities at different heights along the radius in shaft furnace
(a) $h=15m$; (b) $h=6m$

The particle size segregation under different burden profiles in shaft furnace with CGD is shown in Fig. 5-22. In this work, the gray, blue, green and red colors are used to represent pellet, sinter, coke and flux, respectively. There are obvious segregation phenomena for the particles under the 'V' shape, inverse 'V' shape, 'W' shape and 'M' shape burden profile. All these cases show that the bright color particles gather near the burden foot, which indicates the large coke particles tend to roll away from the landing point towards. The main reason is that coke and ore have larger particle sizes, and especially coke has the smallest density among all the burden types. This phenomenon is consist with the reported work, who reported that the large particles tend to move away from the landing point. The burden distribution in shaft furnace under 'Flap' shape profile is shown in Fig. 5-22(e). The particles present a well-mixed state in the furnace, and during the descending process, the particle distribution changes little. A quantitative discussion about the particle segregation will be given later using a particle size segregation index.

Since the particle basically descend along the streamline during the discharging process, the change of particle segregation along the height is not obvious as that in the radial direction. In this work, the height of 15m above the bottom of the furnace is the

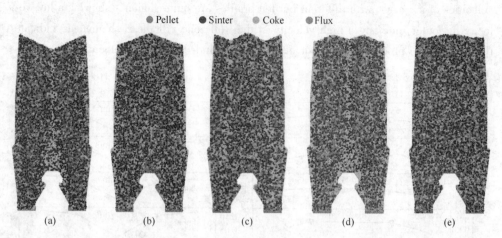

Fig. 5-22 Influence of burden profile on the particle distribution in COREX shaft furnace with CGD
(a) 'V' shape; (b) Inverse 'V' shape; (c) 'W' shape; (d) 'M' shape; (e) 'Flap' shape

main reaction area of the gas-solid phases, which can represent the gas-solid flow-transfer-reaction in the middle upper part of the shaft furnace.

The effect of burden profile on the segregation index in the radial direction is shown in Fig. 5-23. In the case of 'V' shape, the SI_P (P represents pellet) first increases and then slightly decreases from the center to the wall zone, while the SI_C (C represents coke) shows the opposite tendency. The pellet has a negative segregation in the shaft center zone, while the SI_P shows a positive value in the area near the wall zone. On the contrary, the SI_C has a positive value in the center area and a negative value in near the wall zone. This reflects that the spherical small pellets are tend to be near the landing site, while the larger coke particles tend to roll away towards the center area. Due to the

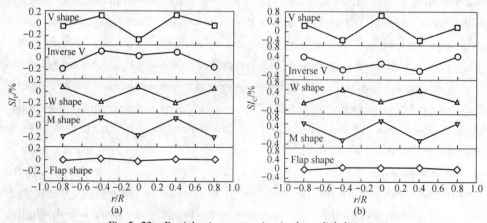

Fig. 5-23 Particle size segregation in the radial direction
(a) Pellet; (b) Coke

opposite pattern of the burden profile, the trend of the SI in shaft furnace under the inverse 'V' shape shows a contrary tendency to that of the 'V' shape burden profile.

For the 'W' shape burden profile, the SI_P is negative in the middle zone of the furnace and it changes to positive in the center and wall area of the furnace. Correspondingly, the variation trend of SI_C presents an M-type in the radial direction. It can be expected that the SI_P and SI_C in shaft furnace under the 'M' shape burden profile have the opposite evolution law to that under 'W' shape burden profile condition. The distribution of SI_P and SI_C in shaft furnace under the 'Flap' shape shows that both the curves fluctuating in a small range near 0, which means that there is no obvious segregation in furnace under 'Flap' shape condition.

Fig. 5-24 shows the influence of burden profile on the average voidage in COREX shaft furnace with CGD. It can be observed that the average void in the packed bed of the shaft furnace presenting somewhat difference. The contour distribution of the void shows that the dark blue area in shaft furnace under 'Flap' shape is somewhat higher than that in other burden profiles, indicating that the overall void is lower. The size-grading caused by particle segregation will make the voids of packed bed under mixed charging smaller than those under the non-flap burden profiles. Besides, all cases show that the largest void fraction occurs near the wall zone due the 'wall effect'. In terms of the overall average porosity, the overall maximum and minimum porosity in the packed bed are in the case of 'V' shape and 'Flap' shape profile respectively. The second highest porosity can be found in the case of inverse 'V' shape profile. Different burden profiles causes different segregation behaviors, which on one hand reflects the change of the particle size distribution in shaft furnace, and on the other hand it also affect the po-

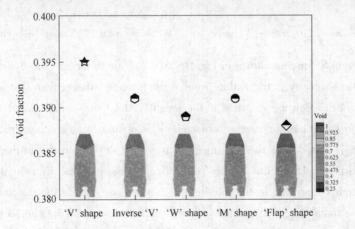

Fig. 5-24 Influence of burden profile on the average void distribution in COREX shaft furnace with CGD

rosity distribution. The next section will discuss the influence of burden profile on the gas flow and pressure distribution in shaft furnace.

5.4.3 Gas Flow and Pressure Distribution

Fig. 5-25 presents the effect of burden profiles on the streamlines of gas flow. As the reducing gas can be injected from both slots and CGD inlets of the furnace, the central gas distribution in the lower part of the shaft furnace can be enriched. The gas flow patterns in the shaft furnace under the five burden profiles are relatively similar. The gas radial distribution at different height is discussed for the further investigation of gas flow in the shaft furnace.

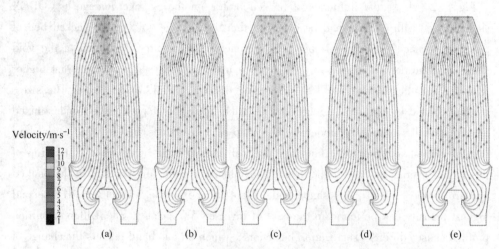

Fig. 5-25 Influence of burden profile on the streamlines of the gas flow in COREX shaft furnace with CGD
(a) 'V' shape; (b) Inverse 'V' shape; (c) 'W' shape; (d) 'M' shape; (e) 'Flap' shape

At the height 5.5m, as shown in Fig. 5-26(a), due to the gas supply from the CGD inlets, the gas velocity in the middle zone of the furnace is faster than that near the wall and center zone. When the gas rises to the height 8.0m, the overall gas velocity increases, and the gas velocity near the furnace wall is obviously larger than that in the furnace center. This is due to the gas injection from the slots. With the further rise of gas flow, at the height 15m, due to the 'wall effect', the gas velocity near the wall zone is slightly higher than that in the middle and center zone, while the gas velocity distribution in the radial direction is more uniform than that in other heights. It can be seen from Fig. 5-26(a), (b) and (c), with the increase of the height, the gas velocity in the furnace under the five burden profiles basically shows an increasing trend. In addition, the mean square deviation of gas velocity in furnace under different burden profiles decreases with the increase of the height, which means the gas distribution in furnace gradually tends to be uniform during the rising process.

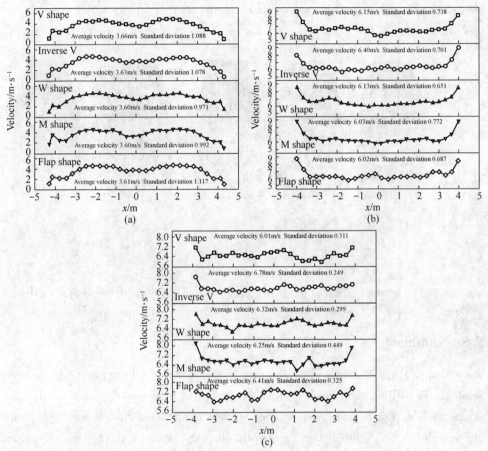

Fig. 5-26 Radial distribution of gas velocity at different height
(a) 5.5m; (b) 8.0m; (c) 15m

Fig. 5-27 shows the effect of burden profile on pressure drop in COREX shaft furnace with CGD. It can be seen that the isobars in the area above the slot of shaft furnace are basically horizontal under five burden profiles. The 3600Pa isobar of the shaft furnace under the 'V' shape and inverse 'V' shape is close to the slot level, while it moves upward under the 'W' and 'M' shape. In the case of 'Flap' shape profile, the 3600Pa isobar is raised by 1.7m relative to the slot level. In the lower part of shaft furnace, the 'V' shape and inverse 'V' shape conditions are mainly 4800Pa isobaric zone. However, in the lower part of furnace under 'Flap' shape, there is a large amount of dark yellow area higher than 4800Pa. Comparison of the pressure drops in whole shaft furnace, the case with 'V' shape has the lowest pressure drop, 5990Pa, while the pressure drop in the furnace under 'Flap' shape is the largest among the five burden profiles, which is 6530Pa. In practical production, if the pressure drop is considered emphatically, the 'V' shape burden profile is given priority during the charging process.

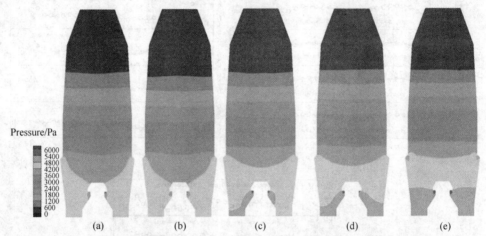

Fig. 5-27 Influence of burden profile on pressure drop in COREX shaft furnace with CGD
(a) 'V' shape; (b) Inverse 'V' shape; (c) 'W' shape; (d) 'M' shape; (e) 'Flap' shape

5.5 Summary

A coupled CFD-DEM model is developed to investigate the influence of CGD on gas-solid flow in COREX shaft furnace. The effect of burden profile on gas-solid distribution is also studied. It principally pays attention on the exploration of particle velocity, particle segregation, void distribution, gas distribution and pressure drop. The findings are summarized below.

The descending velocity of particles located in the upper of furnace is almost unaffected with the installation of CGD. However, the particles were hindered by the CGD with the descending process, especially the burden in the middle area of the furnace, resulting in the lower velocity compared to the case without CGD. The void distribution is affected by the CGD slightly. Meanwhile, the effect of CGD on the gas velocity is obvious, especially below the slot zone; the installment of CGD improves the gas velocity and promotes the center gas distribution. What's more, the CGD can increase the pressure drop in the furnace. Though the appliance of CGD increases the pressure drop partly, the center gas flow was developed significantly with the little influence on the solid flow behavior, which can further promote to obtain the higher metallization rate and gas utilization ratio.

The average residence time of gas phase increased from 4.48s to 4.6s when CGD was installed. The dispersed plug volume fraction and the well-mixed volume fraction increased, while the dead volume fraction decreased in the case with CGD. For the solid phase, the flow was almost entirely composed of the plug flow in the shaft furnace with or without the CGD. Compared with the case without the CGD, the average residence time and the dispersed plug volume fraction decreased in the case with the CGD, while the dead volume fraction and the well-mixed volume fraction increased.

The burden profile has a minor effect on the burden descending velocity in COREX shaft furnace. The distribution of solid velocity under different burden profiles are basically the same. In the area above the slot level, the burden basically descends evenly. In the case of 'V' shape profile, the large particles such as coke are likely to roll away from the landing site, while the pellet is more likely to gather near the burden apex. The segregation phenomenon in furnace with inverse 'V' shape profile is opposite to that in 'V' shape condition. The pellet shows positive segregation near the center and wall area of shaft furnace under 'W' shape profile, while it is opposite in the case of 'M' shape profile. There is no obvious segregation in furnace under 'Flap' shape condition. The overall maximum and minimum porosity in the packed bed are in the case of 'V' shape and 'Flap' shape profile respectively. The second highest porosity can be found in the case of inverse 'V' shape profile. The mean square deviation of gas velocity in furnace under 'W' shape is relatively small. Considering the uniform descending behavior of solid in shaft furnace, the 'W' shaped burden profile can be practiced in actual production to achieve the uniform of gas-solid flow in shaft furnace with CGD. The furnace with 'V' shape profile has the lowest pressure drop, while the pressure drop in the furnace under 'Flap' shape is the largest among the five burden profiles. When considering the pressure drop in practical production, the 'V' shape burden profile is given priority during the charging process.

References

[1] Natsui S, Uder S, Nogami H, et al. Penetration effect of injected gas at shaft gas injection in blast furnace analyzed by hybrid model of DEM-CFD [J]. ISIJ International, 2011, 51, 1410~1417.

[2] Zhou Z Y, Zhu H P, Wright B, et al. Gas-solid flow in an ironmaking blast furnace-II: Discrete particle simulation [J]. Powder Technology, 2011, 208, 72~85.

[3] Yang W J, Zhou Z Y, Yu A B. Discrete particle simulation of solid flow in a three-dimensional blast furnace sector model [J]. Chemical Engineering Science, 2015, 278, 339~352.

[4] Matsuhashi S, Kurosawa H, Natsui S, et al. Evaluation of coke mixed charging based on packed

bed structure and gas permeability changes in blast furnace by DEM-CFD model [J]. ISIJ International, 2012, 52, 1990~1999.
[5] You Y, Luo Z G, Zou Z S, et al. Numerical study on mixed charging process and gas-solid flow in COREX melter gasifier [J]. Powder Technology, 2020, 361, 274~282.
[6] Bai M H, Han S F, Zhang W Y, et al. Long, Influence of bed conditions on gas flow in the COREX shaft furnace by DEM-CFD modelling [J]. Ironmaking Steelmaking, 2017, 44, 685~691.
[7] Hou Q F, Li J, Yu A B. CFD-DEM study of heat transfer in the reduction shaft of COREX [J]. Steel Research International, 2015, 86, 626~635.
[8] Standish N. Studies of size segregation in filling and emptying a hopper [J]. Powder Technology, 1985, 45, 43~56.

6 CFD Simulation of Inner Characteristics in COREX Shaft Furnace with Center Gas Distribution Device

The gas-solid countercurrent moving bed is a common reactor in many industrial processes, such as ironmaking, coal gasification, waste heat recovery, chemical catalysis and other industries. It enables efficient gas-solid flow-transfer-reaction process. The packed bed structure, such as void distribution and particle diameter distribution, directly affects the gas flow, and further determines the heat and mass transfer performances in the reactor. The COREX shaft furnace is a typical counter-current ironmaking reactor. Various charging matrices in the top of COREX shaft furnace may lead to different burden shapes, resulting in different behavior of multi-size particles percolation, which in turn affects the packed bed structure. Appropriate burden profile in shaft furnace may benefit to improve the efficiency of gas-solid transport, thereby increasing the solid metallization ratio (MR) in shaft furnace and reducing the energy consumption in the COREX process. Therefore, it is necessary to study the inner characteristics in COREX shaft furnace under different packed bed structures for control and optimization of the operation.

Due to the 'black boxes' principle of the COREX shaft furnace, it is difficult to obtain the inner multi-phase and multi-field distribution through actual measurement. Numerical simulation methods have become the main method to solve this problem. For example, Dong established a two-dimensional mathematical model of the reduction shaft furnace to study the flow, heat transfer, mass transfer, and reaction phenomenon. Xu took the shaft furnace as the research object and conducted in-depth theoretical analysis and research on the phenomenon of gas flow, heat transfer, and chemical reaction behavior in the shaft furnace. Wu et al. developed a two-dimensional and three-dimensional mathematical model to study the characteristics inside COREX shaft furnace and its operation parameters optimization. Zhang et al. developed steady-state models to analyze the influence of the central gas distribution (CGD) on the inner characteristics of the COREX shaft furnace. Some work focused on the chemical and thermal phenomena in MIDREX shaft furnace. Nevertheless, most works were studied with a constant void and uniform diameter distribution in shaft furnace, and the information about the influence of packed structure, such as void distribution and particle diameter distribution, on the heat and mass transfer are rare.

The combined computational fluid dynamic (CFD) and discrete element method (DEM) is an effective tool to study the micro-scopic and macro-scopic properties of gas-solid behavior in a shaft furnace. However, these models did not take the heat and mass transfer or chemical reactions between gas and solid into account. In particular, due to the limitation of computing resources, the particle diameters in the CFD-DEM model are enlarged.

Therefore, in this chapter, a two-dimensional steady-state mathematical model of the COREX shaft furnace with CGD was developed. The influence of packed structures on the gas flow, pressure distribution, species concentration, and solid metallization ratio in COREX shaft furnace was investigated. The findings of this work can provide a convenient and cost-effective tool to stabilize or improve the furnace performance.

6.1 Mathematical Modelling

6.1.1 Governing Equations

Based on the characteristics of the Euler two-phase flow model, both gas and solid phases are assumed to be fluids, and the calculation is completed by the Navier-Stokes equation. The mass, momentum, energy and species transfer can be described by Eq. (6-1) under steady state.

$$\nabla \cdot (\varepsilon_p \cdot \rho_p \cdot \phi \cdot \vec{v}_p) = \nabla \cdot (\varepsilon_p \cdot D_\phi \cdot \nabla(\phi)) + S_\phi \qquad (6-1)$$

where, the subscript p represents the phase, and D and S represent the diffusion coefficient and source term, respectively. The detailed representation of all variables ϕ is consistent with the previous literature.

According to the gas-solid phase composition and reaction conditions in the COREX shaft furnace, the chemical reactions considered in the model include CO and H_2 reduction of iron oxide. The related reactions and reaction rate expressions are listed in Table 6-1. The chemical reaction rates are calculated by a three-interface unreacted core model based on physical chemistry data from Perry et al.'s book. The reaction rate constants and the effective diffusion coefficients are taken from previous works.

Table 6-1 Chemical reactions and rate expressions in model

n	Reaction	Rate expression
R_1	$3Fe_2O_3+CO \rightarrow 2Fe_3O_4+CO_2$	
R_2	$Fe_3O_4+CO \rightarrow 3FeO+CO_2$	$R_n = \dfrac{6\varepsilon_s}{\phi_s \cdot d_s} \cdot \dfrac{\rho_g}{W \cdot M_g} \sum_{m=1}^{3}[\alpha_{n,m}(Y_{CO}-Y^e_{m\_CO})]$
R_3	$FeO+CO \rightarrow Fe+CO_2$	
R_4	$3Fe_2O_3+H_2 \rightarrow 2Fe_3O_4+H_2O$	
R_5	$Fe_3O_4+H_2 \rightarrow 3FeO+H_2O$	$R_n = \dfrac{6\varepsilon_s}{\phi_s \cdot d_s} \cdot \dfrac{\rho_g}{W \cdot M_g} \sum_{m=4}^{6}[\alpha_{n,m}(Y_{H_2}-Y^e_{m\_H_2})]$
R_6	$FeO+H_2 \rightarrow Fe+H_2O$	
R_7	$CO+H_2O \rightarrow CO_2+H_2$	$R_7 = \dfrac{1000}{(101325)^2} \cdot \varepsilon_s \cdot k_7 \cdot \left(P_{CO} \cdot P_{H_2O} - \dfrac{P_{CO_2} \cdot P_{H_2}}{K_7}\right)$

In the table, ϕ_s is the solid charge shape factor, d_s is the solid charge particle diameter, M_g is the average mass of reducing gas, kg/mol, Y_i is the gas mole fraction, Y^e_i is the gas mole fraction under a certain equilibrium condition.

6.1.2 Boundary Conditions

Fig. 6-1 shows the schematic diagram of COREX shaft furnace with CGD. The reducing gas is blasted into the furnace by the slot gas inlet locked in the bustle zone and the CGD inlet at the bottom. After reacting with the descending burdens, the gas discharged at the top. The burden is charged from the top of the furnace continuously, then reduced to DRI, and finally discharged by screws at the bottom of the furnace. The reducing gas flow is 214000m³/h (standard state), and the flow rates of slot gas inlet and CGD inlet are both 50% of the total flow rate. The volume fractions of CO, CO_2, H_2, and H_2O in the reducing gas are 64.7%, 9.1%, 22.7% and 3.5%, respectively. The gas temperature is 1100K. For the solid phase, the top is the velocity inlet boundary, and the bottom is the outflow. The wall boundaries are considered as the no-slip condition, and the heat transfer coefficient and the temperature of the wall are set as 20W/(m² · K) and 300K, respectively. The density, viscosity, and thermal conductivity the burden are 2273kg/m³, 6kg/(m · s), and 0.8W/(m · K), respectively.

In the practical production process, various charging matrices will lead to different burden surface profiles, which in turn affect the porosity and particle size distribution of the packed bed in shaft furnace. In this work, five packed bed structures corresponding to different burden profiles are studied. The case-Flap, case-InV, case-V, case-M, and case-W are the packed bed structures with 'Flap' shape, inverse 'V' shape, 'V' shape, 'M' shape, and 'W' shape burden profile respectively. The packing density and diameter distribution in the COREX shaft furnace with CGD are shown in

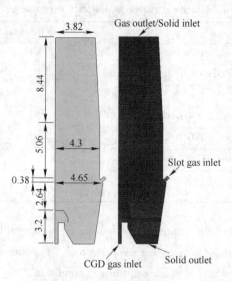

Fig. 6-1 Geometrical structure and mesh of COREX shaft furnace with CGD

Fig. 6-2. The case-Flap is the packed bed with a flap burden profile, and the packed density and solid diameter are uniformly distributed in the furnace. The packing density and average diameter of solid are 0.6 and 15.4mm respectively. Considering that after the burden reaches the landing point, the larger particles are easier to roll, which will result in a smaller particle size and a larger packing density at the tip. Moreover, with the downward movement of burden, the particle size decreases gradually due to the pulverization caused by the burden load and reduction process. Thus, for the case-InV and case-V, at half of the overall height of the packed bed, in the horizontal direction from the packed bed tip to the foot, the packing density decreases from 0.61 to 0.59, and the particle diameter increases from 13.4mm to 17.4mm. For case-M, the pile tip is in the middle position, so at half the height of the packed bed, the packing density decreases from 0.605 at the tip zone to 0.595 at the center and wall area, and the particle diameter increases from 14.4mm to 16.4mm. Case-W shows the opposite of the trend of case-M. In the height direction, along the direction from the top to the bottom of the furnace, the packing density increases by 0.04, and the particle diameter decreases by 2mm.

The conservation equations are solved numerically by the finite volume method with commercial software ANSYS FLUENT (release 19.0). The first order upwind scheme is used for discretization and then the coupled SIMPLE method is applied. The simulation is considered to have converged when the residuals for each variable are less than 1×10^{-5}.

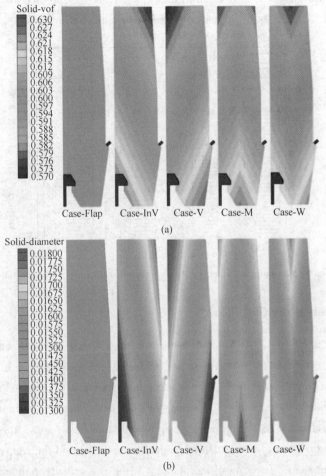

Fig. 6-2 Packing density and diameter distribution in the COREX shaft furnace
(a) Pack density; (b) Diameter distribution

6.2 Results and Discussion

6.2.1 Model Validation

The validation of this mathematical model is mainly based on the top gas composition and the solid metallization rate. Fig. 6-3 shows the comparison between the calculation results and the practical data. It can be seen that the measured results of the top gas composition and solid metallization ratio are in good agreement with the calculated results. The maximum relative error is 5.5%, which is the volume fraction of H_2 at the top of the shaft furnace. The minimum relative error is 1.0%, which is the solid metallization ratio. Although there is a certain error between the measured values and the calculated data, the model is still reliable and can be used to predict the transport phenomena in the COREX shaft furnace with different packed structures.

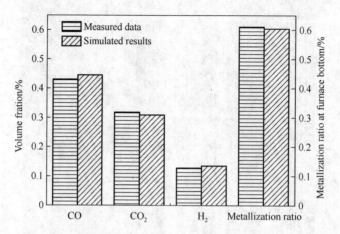

Fig. 6-3 Comparison between the simulated results and measured data

6.2.2 Influence on Gas Flow

The gas distribution in COREX shaft furnace with different packed bed structures is shown in Fig. 6-4. Since 50% reducing gas is blasted into shaft furnace from the slot inlet and the CGD inlet respectively, there is the maximum gas velocity near the inlet area. In the area below the slot inlet, the influence of packed bed structure on the gas distribution is not obvious. As the reducing gas gradually moves upwards, the adjustment of the gas flow by the packed bed structure gradually appears, and the gas velocity contour takes a different shape. The overall morphology of the gas velocity contour on the upper part of the shaft furnace is similar to the distribution of solid packing density. For example, the gas velocity is relatively uniform along the radial direction under the case-Flap condition. In case-InV and case-V, the gas velocity gradually increases and decreases from the center to the edge, respectively. The case-V can promote the development of central gas flow. The velocity in case-M and case-W presents the opposite distribution state. The velocity contour shows M shape in case-M, and in case-W the velocity contour shows a W shape. In areas where the solid packing density is smaller and the particle diameter is larger, the gas flow channel is larger, which promotes the development of gas flow.

Fig. 6-5 shows the influence of packed bed structure on the radial velocity at different heights. In this work, $x=0$ is the central area of the furnace, and $x=4.3$ is the wall zone of the furnace. It can be seen that the radial gas velocity at $y=15.5$m and 10m

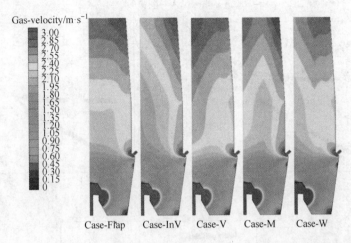

Fig. 6-4 The velocity contour in shaft furnace under different packed structures

show a similar tendency. The gas velocity at height of $y = 15.5$m is slightly higher than that at height of $y = 10$m. This is due to the gradual decrease in the radius of the shaft furnace. The following will mainly analyze the gas velocity under different packed structures at $y = 10$m. It can be seen from Fig. 6-5(b) that the velocity is relatively uniform along the radial direction under the case-Flap condition, and the maximum velocity is 2.35m/s at the wall zone, which is slightly higher than that at the center. The average velocity is 2.267m/s, and the standard deviation is 0.043m/s in case-Flap. Under the condition of case-InV, the gas velocity increased from 1.95m/s to 2.46m/s from the center to the wall zone, and the average velocity and the standard deviation were 2.182m/s and 0.156m/s respectively. The case-V has a larger center gas velocity and a lower wall velocity. The overall average velocity is 2.357m/s, and the velocity deviation is 0.086m/s in case-V. For case-M, the velocity in the middle zone is lower, while has a larger velocity at the center and wall zone of the furnace. The minimum velocity in the middle zone is 2.135m/s, and the radial average velocity is 2.267m/s, the speed deviation is 0.092m/s. Case-W has the highest velocity in the middle zone, which can reach 2.342m/s. The average velocity is 2.27m/s, and the standard deviation is 0.057m/s. In terms of even distribution of gas phase, the order is case-Flap>case-W>case-V>case-M>case-InV.

Fig. 6-6 shows the pressure distribution in shaft furnace under different packed structures. It can be seen that along the height direction from the top to the bottom, the pressure increases gradually under different packed bed structures. These pressure distribution curves present a three-stage distribution characteristic. One is in the area

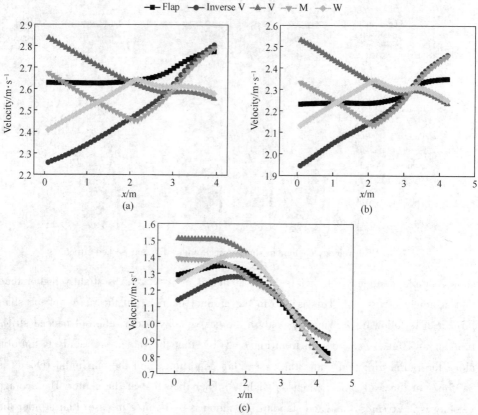

Fig. 6-5 Radial gas velocity at different heights
(a) $y=15.5m$; (b) $y=10m$; (c) $y=3m$

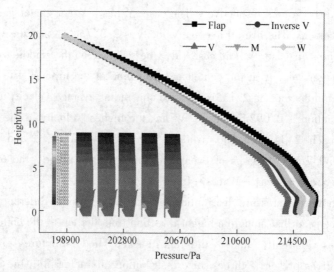

Fig. 6-6 Pressure distribution in shaft furnace under different packed structures

above the slot inlet with a height of 5.84m, the curves show a uniform decrease with the increase of height. The second is that when the height is lower than 5.84m, the curves have a turning point, and the pressure change range decreases along with the height; The third is that after the height is reduced to the area below the CGD inlet, the pressure remains steady. For different packed bed structures, the order of total pressure drop in shaft furnace is case-Flap>case-M>case-W>case-InV>case-V.

6.2.3 Influence on Gas and Solid Composition

The mole fraction distributions of CO and CO_2 in the COREX shaft furnace under different packed bed structures are shown in Fig. 6-7. It can be seen that there is the highest CO concentration at the slot inlet and CGD inlet zone. As the gas diffuses outward, CO participates in the reduction reaction. According to the three-interface unreacted core model, the reduction reaction of iron ore is an equal-molar reaction to the gas phase. Thus, the changing trend of CO_2 is exactly the opposite of that of CO. For differ-

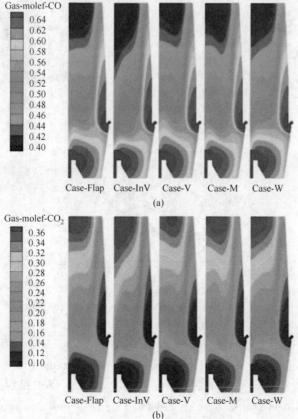

Fig. 6-7　CO and CO_2 distribution in shaft furnace under different packed structures

ent packed bed structures, the mole fraction distribution of CO and CO_2 presents different characteristics. In terms of conditions conducive to the development of edge gas, such as the case-InV, the development of the center gas is insufficient, so the CO concentration of center gas is lower and the CO_2 concentration is higher. Fig. 6-8 shows the radial species distribution at different heights. It can be seen from Fig. 6-8(a) and Fig. 6-8(b), the CO concentration under different packed bed structures has a low value in the center zone and a large value near the wall zone. Under each packed bed structure, the CO concentration in the central area from high to low is case-V> case-M> case-Flap> case-W> case-InV. The case-V is conducive to promoting the development of the central gas distribution, increasing the reduction potential in the central zone, and promoting the uniform distribution of the metallization rate in the radial direction. At $y=3m$, the gas mole fraction is mainly affected by the gas flow at the CGD inlet, and the CO concentration in the center of the shaft furnace is higher.

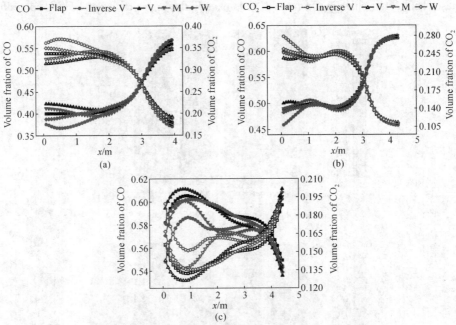

Fig. 6-8 Radial species distribution at different heights
(a) $y=15.5m$; (b) $y=10m$; (c) $y=3m$

Fig. 6-9 shows the FeO and Fe mole fraction distribution under different packed bed structures. The reduction of iron ore follows the gradual transformation principle, that is, the iron ore changes sequentially from $Fe_2O_3 \rightarrow Fe_3O_4 \rightarrow FeO \rightarrow Fe$. According to the thermodynamic characteristics of the six reduction reactions shown in Table 6-1, the equilibrium constants K_1 and K_4 are much larger than the other equilibrium constants. Therefore, when the burden is added from the top of the furnace, the ore is quickly reduced from Fe_2O_3 to Fe_3O_4. With the movedown of the solid phase, the burden is gradually reduced from Fe_3O_4 to FeO. However, the equilibrium constants K_3 and K_6 are much smaller than other equilibrium constants. Therefore, the reaction conditions for the reduction of FeO by CO and H_2 are relatively harsh, and the progress of the reaction requires a higher temperature and gas concentration. In the area near the slot inlet and the CGD inlet, the stepwise reduction of the upper iron oxide and the higher reduction potential are beneficial to the reduction of FeO. For different packed bed structures, the FeO content in the central area is high. The porosity and particle size distributions in different packed beds affect the reduction of iron. Comparative analysis of the position of the FeO contour, the higher position of FeO, the easier it is to reduce to FeO and further reduce to Fe. One can observe that the bottom of the FeO contour under the case-InV condition is lower, while the case-V is the highest. This characteristic presents the final metallization rate in the furnace.

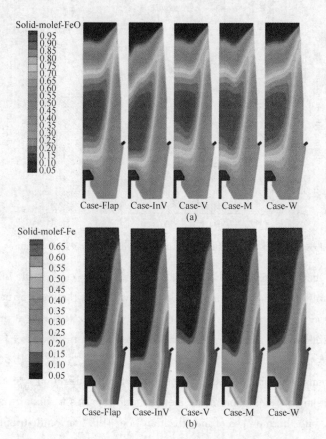

Fig. 6-9 FeO and Fe distribution in shaft furnace under different packed structures

Fig. 6-10 shows the solid metallization rate under different packed bed structures. The metallization rate is an important index for shaft furnace production. A higher metallization rate means that the furnace is in an efficient and stable state, which is beneficial to reduce the fuel ratio in the melter gasifier. The metallization rate is determined by the solid phase composition, as shown in Eq. (6-2). With the solid moving downward, more and more FeO is reduced to Fe.

$$\text{Solid MR} = \frac{w_{Fe}}{\frac{112}{160} \cdot w_{Fe_2O_3} + \frac{168}{232} \cdot w_{Fe_3O_4} + \frac{56}{72} \cdot w_{FeO} + w_{Fe}} \tag{6-2}$$

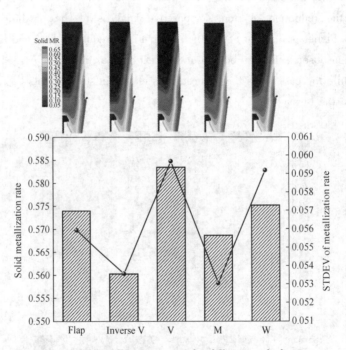

Fig. 6-10 Solid metallization rate under different packed structures

It can be seen that the low-metallization rate area with blue color is the highest in case-V, reflecting the larger metallization rate in the low part of the shaft furnace, while the blue area in case-InV is the lowest. For the average metallization rate at the bottom, the average metallization rate from large to small is case-V> case-W> case-Flap> case-M> case-InV, and the metallization rates are 0.583, 0.575, 0.574, 0.569, and 0.56 respectively. Comparing the uniform distribution of the metallization rate along the radial direction, the standard deviation of the metallization rate has a similar trend to the metallization rate. With the increase of metallization rate, the uneven distribution along with the radial direction increases.

Based on the above research results, it can be seen that the packing density and particle size distribution in the packed bed of COREX shaft furnace directly affects the gas flow distribution and then further have an impact on the reaction process in the furnace. This influence is ultimately reflected in the difference in the metallization rate of the solid phase. The pressure drop is low, and the metallization rate is the highest in case-V. In the process of pursuing the improvement of the metallization rate, the charging matrix can develop toward the V-shaped packed structure.

6.3 Summary

A two-dimensional steady-state mathematical model is developed in this work to study the influence of packed bed structure on the inner characteristic in COREX shaft furnace with CGD. The gas velocity, pressure distribution, gas and solid composition, and metallization rate are discussed. The conclusions from the present study can be summarized as follows:

(1) The gas distribution in the case-Flap is relatively even along the radial direction. The gas velocity under the case-InV and case-V conditions gradually increase and decrease from the center to the wall zone, respectively. The case-V can promote the development of central gas. The velocity contour of the case-M and case-W show an M and W shape respectively. For the even distribution of the gas phase, the order is case-Flap> case-W> case-V> case-M> case-InV.

(2) The pressure distribution presents a three-stage distribution characteristic. For different packed bed structures, the overall pressure drop under the five packed bed structures is case-Flap> case-M> case-W> case-InV> case-V.

(3) The order of average metallization rate from large to small is case-V> case-W> case-Flap> case-M> case-InV. The pressure drop is low, and the metallization rate is the highest in case - V. The V shaped burden profile can be practiced in actual production to achieve the larger metallization rate.

References

[1] Dong X F, Xiao X G, Zou Z S. Mathematical simulation of COREX pre-reduction shaft furnace [J]. Journal of Northeastern University (Nature Science), 1998, 19, 229~232.

[2] Xu H C. Numerical simulation on the process of COREX off gas shaft furnace [M]. Beijing. 2000.

[3] Wu S L, Xu J, Yang S D, et al. Basic characteristics of the shaft furnace of COREX® smelting reduction Process Based on Iron Oxides reduction simulation [J]. ISIJ International, 2010, 50, 1032~1039.

[4] Zhang X S, Luo Z G, Zou Z S. Numerical analysis on the performance of COREX CGD shaft furnace with top gas recycling [J]. ISIJ International, 2019, 59, 1972~1981.

[5] Ghadi A Z, Valipour M S, Biglari M. CFD simulation of two-phase gas-particle flow in the Midrex shaft furnace: The effect of twin gas injection system on the performance of the reactor [J]. International Journal of Hydrogen Energy, 2017, 42, 103~118.

[6] Shams A, Moazeni F. Modeling and simulation of the MIDREX shaft furnace: reduction, transition and cooling zones [J]. JOM, 2015, 67, 2681~2689.

[7] Perry R H, Green D W, Maloney J O. Perry's chemical engineers' handbook [M]. 7th edition. New York: McGraw-Hill, 1997.

[8] Takahashi R, Takahashi Y, Yagi J, et al. Operation and simulation of pressurized shaft furnace for direct reduction [J]. Transactions of the Iron and Steel Institute of Japan, 1986, 26, 765~774.

Part III

Physical and Mathematical Simulation of COREX Melter Gasifier

7 Numerical Simulation of Combustion Characteristics in the Dome Zone of the COREX Melter Gasifier

The dome zone is one of the most significant areas of the COREX melter-gasifier. There is a secondary injection system for keeping stable dome temperature in COREX process, where the combustion of the recycling dust is one of the primary factors affecting dome temperature. Unsuitable dome temperature would lead to many issues, including blocking gas outlet, the erosion of refractory and adherence of particle. Therefore, it is instructive to study the transport phenomena in the dome zone, especially the combustion of the recycling dust.

The recycling dust is composed of unburnt fine coke, coal particle and metallic iron and other nonmetallic oxide and its combustion process can be divided into four stages: the combustion of volatile matter, coal char, fine coke, and the burnout process. However, the effect of the combustion of recycling dust on the coupling fields in the dome needs further research. It is worth noting that some researches about the combustion behavior of other carbonaceous materials, including pulverized coal, charcoal, biomass, provide a new idea for this study. Comparing to redundant industrial experiment, the computational fluid dynamics (CFD) that couple heat, mass and momentum balances is a better choice to explore the combustion process of recycling dust and its influence on internal characteristics in the dome.

Some researchers have done some works on combustion simulation of dust. For example, Berger et al. firstly studied the combustion behavior in the dome of melter-gasifier through numerical simulation. But the imperfect model leads to the issue that the predicted average dome temperature is 1174K, which is lower than the minimum temperature in actual production. Du et al. developed a three-dimensional (3-D) mathematical model and concluded the effect of oxygen flows on dome temperature. Similarly, an improved model was developed by Wang et al. to investigate the effect of dust combustion with different granularities on the temperature and gas distribution in the dome. Among which, excessively high volatile content in dust composition may be the main reason for the discrepancy between simulation results and industrial data. The researches mentioned above provide great help for the framework construction of this work. However, the effect of the characteristic of gas flow that coming from packed bed on coupled-field in the dome cannot be ignored, which do not have many discussions in their results.

In fact, the features of rising gas mainly depend on burden structure and the hearth state in lower part of melter-gasifier. For reducing the production cost, some enterprises sometimes add residual inferior coke or semi-coke from blast furnace smelting process to the melter-gasifier as a substitute of high-quality lump coal. But such measures can reduce gas flows by 20% ~ 30%, which will result in extremely low metallization rate in shaft furnace. In addition, the coal rank and particle size are also the primary factors affecting the gas yield of lump coal. In general, strengthening smelting is an effective measure to reduce coke ratio, for example, the injection of pulverized coal or coke oven gas (COG) from the tuyere in COREX melter-gasifier as shown in Fig. 7-1. But these measures will cause shortage of heat in the hearth, so the temperature and compo-

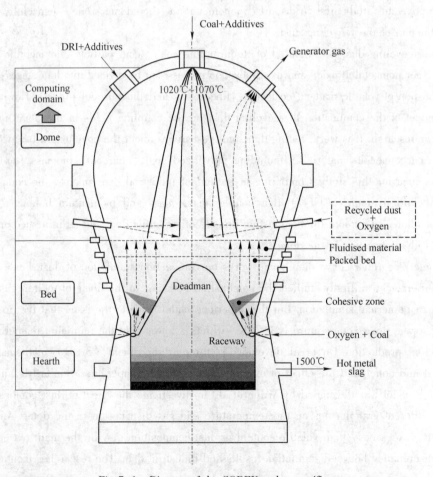

Fig. 7-1 Diagram of the COREX melter-gasifier

nent of rising gas will also change. Once the characteristics of rising gas change, the coupled-field in the dome of melter-gasifier must also be affected. However, there is limited information about the effect of rising gas on the inner features in the dome zone.

To fulfil the research gap, a 3-D steady-state mathematical model to study the transport phenomena in the dome zone of the COREX melter-gasifier. The distributions of velocity, temperature and gas composition in the dome are described, and the conversion behavior and particle trajectories of recycling dust under the local oxygen-rich condition are analyzed. The influences of key operating parameters, such as flow rate, component and temperature of rising gas, on the dome temperature and gas distribution are discussed. The results are expected to tackle resource adjustment challenges and provide guidance for the optimization of the COREX process.

7.1 Mathematical Modelling

7.1.1 Governing Equations

The gas phase is treated as a continuous phase with the assumption of an incompressible ideal gas. Governing equations in terms of mass, momentum, energy, turbulent kinetic energy, turbulent dissipation rate, and species are solved for the gas phase, as documented in the literature. Due to the more accurate prediction to round jets, the gas turbulence is resolved by a realizable $k-\varepsilon$ turbulence model. As shown in Fig. 7-2, when the realizable $k-\varepsilon$ model is adopted, the turbulent viscosity distribution is more uniform.

The governing equations for the gas phase are summarized in Table 7-1.

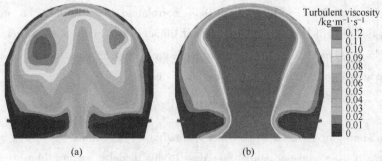

Fig. 7-2 The turbulent viscosity distributions simulated by different turbulence models

(a) The standard $k-\varepsilon$ model; (b) The realizable $k-\varepsilon$ model

Table 7-1 Governing equations for gas phase

Continuity	$\nabla \cdot (\rho U) = \sum_{n_p} \dot{m}$
Momentum	$\nabla \cdot (\rho U U) - \nabla \cdot [(\mu + \mu_t)(\nabla U + (\nabla U)^T)] = -\nabla(p + \frac{2}{3}\rho \tau) + \sum_{n_p} f_D$
Energy	$\nabla \cdot \left[\rho U H - \left(\frac{\lambda}{Cp} + \frac{\mu_t}{\sigma_H}\right)\nabla H\right] = \sum_{n_p} q$
Turbulent kinetic energy	$\nabla \cdot \left[\rho U k - \left(\mu + \frac{\mu_t}{\sigma_k}\right)\nabla k\right] = (P_k - \rho \varepsilon)$
Turbulent dissipation rate	$\nabla \cdot \left[\rho U \varepsilon - \left(\mu + \frac{\mu_t}{\sigma_k}\right)\nabla \varepsilon\right] = \frac{\varepsilon}{k}(C_1 P_k - C_2 \rho \varepsilon)$
Gas mass fraction	$\nabla \cdot \left[\rho U Y_i - \left(\Gamma_i + \frac{\mu_t}{\sigma Y_i}\right)\nabla Y_i\right] = W_i$

$\mu_t = C_\mu \rho \frac{k^2}{\varepsilon}$; $P_k = (\mu + \mu_t)\nabla U \cdot [\nabla U + (\nabla U)^T]$; $i = O_2, CO, CO_2, H_2, H_2O$

The recycling dust is treated as a discrete phase. The solid volume fraction is only $2.395 \times 10^{-4}\%$ in this model, so the DPM model is applicable. Due to its insignificant contribution in all forces exerted on each particle, particle-particle interaction force is neglected. The particle motion is calculated by integrating Newton's second law, of which the turbulent dispersion and the drag force (f_D) are considered. Three heat transfer modes, including convective heat transfer, latent heat transfer related to mass transfer, and radiative heat transfer, are considered for particles. The specific energy equation is shown in Table 7-2. The λ is the thermal conductivity of the gas phase, $W \cdot m^{-1} \cdot K^{-1}$, which can be solved by the formula below. The Nusselt number, Nu, is evaluated by the Ranz - Marshall method correlation. In addition, Lu et al. also summarized the prediction method of the heat transfer coefficients for a variety of non-spherical biomass particles, which can provide great help for our study, i.e. the effect of particle shape on the combustion behavior of dust.

$$Nu = \frac{h d_p}{\lambda} = 2.0 + 0.6 Re^{1/2} Pr^{1/3} \qquad (7-1)$$

As simulation process need to consider the radiation heat transfer between gas phase and particle phase, and the optical thickness in this model is greater than 1, so the P1 model is adopted. Among the gas absorption coefficients are evaluated by the weighted sum of the grey gases model (WSGGM). Particles are fully coupled with the gas phase in terms of mass, momentum, and energy exchanges. The governing equations for the particle phase are shown in Table 7-2.

Table 7-2 Governing equations for particle phases

Mass	$\dfrac{dm_p}{dt} = -\dot{m}$
Momentum	$m_p \dfrac{dU_p}{dt} = -f_D = \dfrac{1}{8}\pi d_p^2 \rho C_D |U - U_p|(U - U_p)$
Energy	$m_p C_p \dfrac{dT_p}{dt} = -q = \pi d_p \lambda Nu(T_g - T_p) + \sum \dfrac{dm_p}{dt} H_{\text{reac}} + A_p \varepsilon_p (\pi I - \sigma_B T_p^4)$
	$C_D = \max(24(1 + 0.15Re^{0.687})/Re, 0.44)$

7.1.2 Chemical Reaction Model

The recycling dust is separated from the product gas by a thermal cyclone separator at about 800℃ ~ 850℃, much higher than the evaporation temperature of the water. Thus, the evaporation processes are not considered in this work. The dust particle is regarded as a combination of fixed carbon and ash, and the latter does not participate in any chemical reactions. Once dust particles and oxygen are injected into the dome zone, multiple chemical reactions will occur, including homogeneous reactions and the oxidation/gasification of carbon in the dust.

Due to the consideration of effect of kinetic and turbulence, the Finite Rate/Eddy Dissipation model are widely adopted to simulate intensive turbulent combustion. Zaitri et al. presents that the Finite Rate/Eddy Dissipation model predicts more accurately the flame structure and gives a best in temperature predication. The dome of melter-gasifier is a complex zone where turbulence and chemical reactions interact, so the Finite Rate/Eddy Dissipation model is adopted in simulation. This model combines the Arrhenius formula with eddy dissipation equations and the net reaction rate is determined by the smaller one of them:

$$R_{\text{react}, i, r} = (v''_{ri} - v'_{ri})\left(k_{\text{react}, rf} \prod_{i=1}^{N}[c_i]^{\eta_{i, rf}} - k_{\text{react}, rb} \prod_{i=1}^{N}[c_i]^{\eta_{i, rb}}\right) \quad (7-2)$$

$$R_{\text{react}, i, r} = \min(R_{\text{react}, i, r}^{(\text{reactans})}, R_{\text{react}, i, r}^{(\text{products})}) \quad (7-3)$$

$$R_{\text{react}, i, r}^{(\text{reactants})} = v'_{ri} M_{w, i} A\rho \dfrac{\varepsilon}{k} \min_{R}\left(\dfrac{Y_R}{v'_{rR} M_{i, R}}\right)$$

$$R_{\text{react}, i, r}^{(\text{products})} = v''_{ri} M_{w, i} AB\rho \dfrac{\varepsilon}{k} \min_{R} \dfrac{\sum_P Y_R}{\sum_{i=1}^{N} v''_{ri} M_{w, i}}$$

$$k_r = A_r T^{\beta_r} e^{-E_r/RT} \qquad (7-4)$$

The heterogeneous reactions consist of carbon combustion in dust particles, Boudouard reaction, and water-gas reaction, which are described by multiple surface reactions model. The abovementioned chemical reactions and their reaction kinetics are summarized in Table 7-3. Compared to the exothermic carbon combustion reaction, some reactions with mediate exothermic or endothermic features exist, such as the oxidation reaction of alkali metal. These reactions are neglected for simplifying model implementation. In addition, the specific definition of the symbols mentioned above can be found in the Nomenclature Table.

Table 7-3 Kinetic parameters of chemical reactions

Chemical Reactions	A_r / s^{-1}	β_r	$E_r / J \cdot kmol^{-1}$
$CO + 0.5O_2 \rightarrow CO_2$	2.20×10^{11}	0	1.67×10^8
$H_2 + 0.5O_2 \rightarrow H_2O$	6.80×10^{15}	0	1.51×10^7
$C(s) + 0.5O_2 \rightarrow CO$	1.36×10^6	0.68	1.30×10^8
$C(s) + CO_2 \rightarrow 2CO$	6.78×10^4	0.73	1.63×10^8
$C(s) + H_2O \rightarrow CO + H_2$	8.55×10^4	0.84	1.40×10^8

7.2 Simulation Conditions

7.2.1 Properties of the Recycling Dust

The component of the recycling dust used in this work is obtained from the COREX-3000, a typical smelting reduction process in China. As shown in Table 7-4, TC and TFe account for nearly 80% of the dust, indicating that the recycling dust has a great potential to be a secondary fuel. Except for the fixed carbon, the other components in the recycling dust (e.g., SiO_2, CaO, MnO, MgO) do not participate in any chemical reactions. The dust particle diameter is set to be 28.88μm with a uniform distribution, and the particle density is 2200kg/m³. The specific heat capacity is consistent with the setting of anthracite, 1680J/(kg·K).

Table 7-4 Chemical composition of the recycling dust (mass fraction, %)

TC	TFe	MFe	SiO_2	Al_2O_3	CaO	MgO	MnO	P_2O_5	S
48.47	29.80	9.58	6.81	2.43	4.68	1.17	0.15	0.05	0.46

7.2.2 Geometry and Boundary Conditions

As shown in Fig. 7-3(a), the geometry and dimensions refer to a commercial melter-gasifier and focus on the dome zone. The bottom surface is set as the inlet for the reducing gas. Six oxygen burners (OB) and four dust burners (DB) at the lower part of the geometry. Their distributions and positions in the present study are consistent with that in the literature. Industrial pure oxygen (99.5% O_2) is introduced by cylindrical hollow pipes in OB. Meanwhile, the DB is equipped with a co-axial lance, of which the oxygen transports in an external tube and the dust is injected in the inner tube with a carrier gas (100% N_2). All lances have an inclination angle of 8° with their tips at the center line, as shown in Fig. 7-3(b). The monitoring line is the level line along DB.

Four gas outlet nozzles at the top of the melter-gasifier are used to detect the composition and temperature of gas products, which are set as a pressure outlet. The walls are assumed as slip-free and adiabatic. The detailed operating parameters of the melter-gasifier are summarized in Table 7-5. In order to understand the effect law of several key operation parameters of rising gas on the dome temperature and gas distribution, the specific experiment plan has been made in Table 7-6.

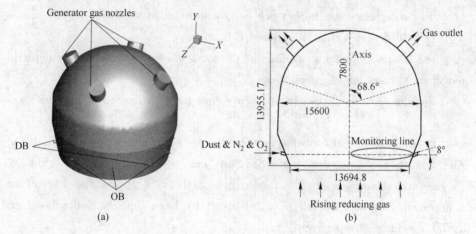

Fig. 7-3 The schematic diagram of the dome zone of the melter-gasifier (Unit: mm)

Table 7-5 The operating parameters of the melter-gasifier

Operating Parameters	Value	Temperature
Generator gas pressure	350 kPa	-
Oxygen supplied by OB (standard state)	3974.84m³/h	298K
Oxygen supplied by DB (standard state)	11803.66m³/h	298K
Nitrogen supplied by DB (standard state)	1700.00m³/h	360K
Recycling dust supplied by DB	246.40kg/min	360K
Rising gas (standard state)	2477.47m³/min	1143K

Continued Table 7-5

Operating Parameters		Value	Temperature
Gas composition of rising gas (volume fraction)	CO	71.1%	
	CO_2	4.8%	
	H_2	18.9%	
	H_2O	2.6%	

Table 7-6 The detailed experiment plan of different operation parameters

Operation parameters	Case 1	Case 2	Case 3	Case 4
Flow rate(v)/m·s^{-1}	0.18	0.28 (base case)	0.38	0.48
$(CO+H_2)(R)$/%	80	85	90(base case)	95
Temperature(T)/K	1043	1093	1143(base case)	1193

Some assumptions are made in the study to improve simulation efficiency, including:

(1) CH_4 is neglected due to its much lower concentration compared to other species (i.e., O_2, N_2, CO, CO_2, H_2, and H_2O).

(2) The composition and temperature of the rising gas are regarded as a uniform diffusion when it enters into the dome zone.

(3) No adherence or agglomeration occurs during the movements of the dust particles.

(4) Dust particles-wall collisions do not influence particle heat and mass transfer.

All simulation cases are performed on the commercial software FLUENT 14.5. Pressure-velocity coupling is treated by a SIMPLE algorithm. The gradient and pressure in spatial discretization were calculated by Least Squares Cell-Based and PRESTO algorithms, respectively.

7.3 Result and Discussion

7.3.1 Model Validation

A mesh-independence test and model validation are conducted before the simulation. Four different mesh resolutions with the cell number of 145669, 195606, 258218, and 349482 are adopted for the mesh independence analysis. Fig. 7-4 depicts the distribution of temperature along with the height of the central axis in the dome with different mesh resolutions. The variation trends of the four curves are almost consistent. The small discrepancies may be attributed to the disparity in the location of the mesh nodes. Thus, to balance numerical accuracy and efficiency, 258218 cells were used in the following simulations.

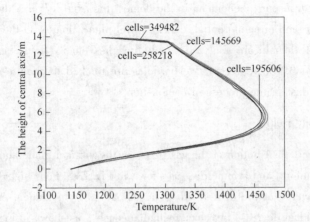

Fig. 7-4   Comparison of the simulation results based on different meshes

The mathematical model is validated in terms of generator gas composition at the outlet and dome average temperature. Practical production presents that the generator gas temperature should be kept from 1273 K to 1343 K. And the proportion of effective components ($CO+H_2$) in generator gas needs to be greater than 80%. As summarized in Table 7-7, the predicted results are in line with the production requirements. Although the simulation process is established based on production conditions, some extremely complicated behaviors are simplified. It is inevitable that there are some small discrepancies appear between the simulation and experimental data. It can be explained that iron oxides in the recycling dust may consume some reducing gas. But these phenomena are not considered in the simulation, so the percentage of effective components in the simulation are slightly higher than that in the experiment. The relative error of the gas temperature at the outlet between the measured data and simulation results is 0.015%, indicating that the current model can reproduce the combustion behavior of recycling dust in the dome zone.

Table 7-7   The comparison between simulated data and measured data

Reducing gas	Measured data	Simulated data	Absolute error	Relative error
Effective components	72.220%	74.195%	1.975%	2.735%
Temperature	1317.0K	1319.0K	2.000K	0.015%

7.3.2 Interpretation of Base Model

To well describe transport phenomena in the dome, the axial (i. e., the vertical plane in Fig. 7-3(b)) and circumferential (i. e., level plane along with the DB) planes of the computational domain are analyzed, including the velocity field, temperature field and gas composition. 300 representative particles are tracked to analyze some variations of the recycling dust in the process of combustion.

7.3.2.1 Velocity Field

The overall velocity distribution of the gas phase in the vertical plane and the circumferential plane is uniform and symmetric, as shown in Fig. 7-5. Four high-speed jets from the DB inlet converge at the central area, accompanied by a gradual expansion in the volume and a gradual decrease in velocity simultaneously, as shown in Fig. 7-5(b). The velocity contour region before OB is much smaller than that before DB due to the lower flow rate and no violent combustion reaction. In addition, there is no low-speed circumfluence formed in Fig. 7-5(a) for the dust particles to maintain a relatively stable motion. Compared with the enormous internal space of the dome, the narrow outlet will sharply increase the velocity of the airflow at the gas generation nozzle, which can reach a maximum value of 6.8m/s.

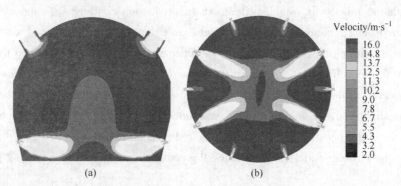

Fig. 7-5 The distribution of velocity field in the dome zone
(a) Vertical plane; (b) Level plane

7.3.2.2 Temperature Field

Fig. 7-6 depicts the temperature distribution in the dome zone. When moving into the front of the dust burner, the high-speed oxygen is fully mixed with the recycling dust particles and reducing gas. Then, intense combustion reactions occur, including homogeneous and heterogeneous reactions, accompanied by a massive heat release. The heat release intensity is greater than the heat transfer intensity during the combustion process, leading to the rapid formation of the burning zone and high-temperature zone in front of DB. The highest temperature reaches more than 2400K. In addition, there is a great difference between the upper and lower parts of the annular flame in Fig. 7-6(a) (i. e., the former has the more pronounced combustion area). A similar phenomenon was also reported in the literature. Because the initial momentum opposes the buoyancy caused by the rising gas fed from the packed bed, the development of the lower part of the flame is interrupted. The momentum at the end of the flame reaches zero essentially. Therefore, the flame bends upwards under the influence of buoyancy. Thus, only the combustion reactions of partial gas take place in front of OB, which play an auxiliary role in regulating temperature in the dome zone. Consequently, the combustion zone before OB is smaller, and no obvious temperature gradient exists in the edge region of the dome zone.

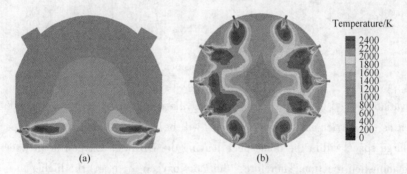

Fig. 7-6 The distribution of temperature field in the dome zone
(a) Vertical plane; (b) Level plane

7.3.2.3 Gas Composition

Apart from heat transfer, the appropriate composition of gas products in the melter-gasifier is also a crucial factor affecting the COREX process. As shown in Fig. 7-7, CO and CO_2, H_2 and H_2O show opposite distributions. The volume fraction of CO and H_2 is the lowest while that of CO_2 and H_2O is the highest in front of DB. This is because the intense combustion reaction results in the massive production of CO_2 and H_2O. Besides, along the monitoring line, there is a local region reaching over 0.75 where the volume fraction of CO is the highest. It may be attributed to the gasification reaction of unburned carbon in the recycling dust that occurs in the environment without oxygen. In the central region of the dome, the volume fraction of CO and H_2 is higher than that in the surrounding region owing to the diffusion effect of rising gas. As the axial height increases, the diffusion effect is more significant due to the space expansion.

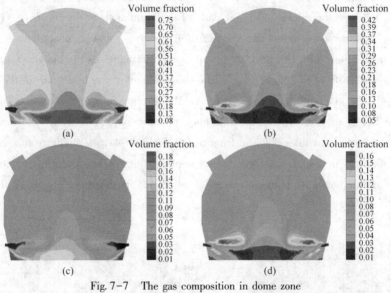

Fig. 7-7 The gas composition in dome zone
(a) CO; (b) CO_2; (c) H_2; (d) H_2O

The variation trend of several variables along the monitoring line is shown in Fig. 7-8. Nitrogen as the carrier gas for dust particles will be rapidly diluted by other types of gas in the large space within the dome once leaving the DB inlet. Oxygen is an integral part of the combustion reaction. Therefore, the more oxygen there is, the higher the reaction temperature. In zone I (0~1.5m), because of the high concentration of O_2, oxidation reactions develop quickly, including the combustion of the reducing gas and dust particles. Therefore, CO_2 and H_2O will have a significant volume fraction increase. As O_2 is gradually consumed, the carbon in the dust burns to CO at low-O_2 areas. After about 2.5m away from the DB, oxygen is almost exhausted. Meanwhile, the temperature reaches a maximum of 2500K. In zone II (1.5~4.0m), the gasification reaction of residual carbon in dust become the dominant response. So the concentration of CO and H_2 keep in up, while the temperature of the gas and volume fraction of CO_2 and H_2O slowly decrease. Finally, there will reach a chemical equilibrium. In zone III (4.0~7.25m), each variable tends to remain constant after undergoing a rapid change, which is attributed to the spreading effect when reducing gas rises. Therefore, it must ensure that the carbon in the recycling dust is completely consumed before reaching zone III, with the aim to reuse resources.

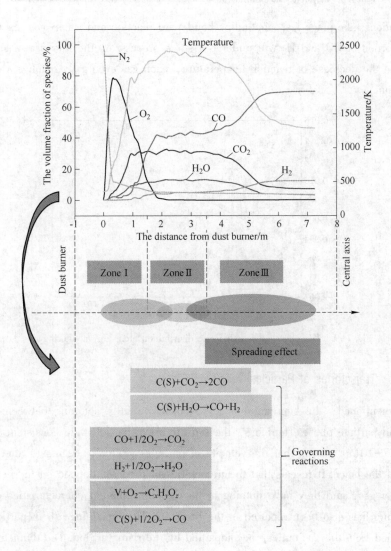

Fig. 7-8 The variation tendency of several variables along the monitoring line

To verify the above analysis, the chemical reaction rate distribution involved in the simulation process is shown in Fig. 7-9. It can be found that this reaction region is composed of multiple overlap reactions. When O_2 enters into the dome zone, it will be consumed by various oxidation reactions. Since CO accounts for more than 60% of the gas in dome zone, it firstly occurs oxidation reaction and has the highest reaction rate. With the rise of temperature, volatiles are released from the dust particles and cracked into hydrocarbons. So the combustion reaction rate of volatile and H_2 increases gradually. Meanwhile, the residual carbon from the dust will also be involved in the oxidation reaction. The gasification reactions take place last, including boudouard reaction and water-gas reaction, so the proportion of CO and H_2 will undergo a surge process. As the increase of spreading effect and the decrease of reaction temperature, each gas eventually reaches a dynamic equilibrium.

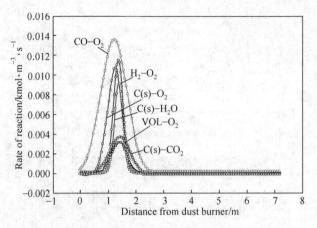

Fig. 7-9　The chemical reaction rate distribution along the monitoring line

7.3.2.4　Trajectories of Particles

As aforementioned, the simulation is conducted with fully coupling between the gas phase and particle phase. Therefore, the trajectories of particles are consistent with the gas. Fig. 7-10(a) shows that the velocity of particles gradually decreases after leaving the tip of the lance. It reveals that the trajectories of particles depend on the velocity of the carrier gas, and they have nothing to do with the inherent characteristics of particles, which has also been reported in the literature. The remaining ash in particles will be expelled from the gas outlet, accompanied by the rising airflow. The diameter of the dust particle first increases and then decreases as it moves along the centre line of the lance, as shown in Fig. 7-10(b). The iron ores and fine coke/coal are primary components in the recycling dust. They both show obvious swelling behavior under a higher heating rate. It can be inferred that the dust particles will also expand after entering into the high-temperature region, potentially increasing porosity simultaneously, improving the diffusion rate of the reactive gas in the particles. After that, as the intense oxidation/gasification reactions proceed, the dust particle diameter decreases rapidly. The dust particle diameter finally keeps constant, meaning that the carbon has been completely consumed. In Fig. 7-10 (c), the burnout rate variation of the dust particle also verifies the conclusion mentioned above. It can be found that once the dust particle enter into the dome zone, it will under a strong combustion process. In the center of the jet flow, there is a volatile enrichment zone, so insufficient oxygen concentration result in a low burnout rate of dust particle. Due to dust particle has a long residence time and a large reaction space in the dome of COREX melter-gasifier, its burnout rate can finally reach more than 99.9%.

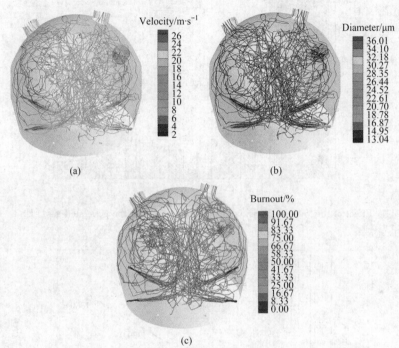

Fig. 7-10 The trajectories of particles coloured by different variables
(a) Velocity; (b) Diameter; (c) Burnout rate

7.3.3 Effect of the Flow Rate of Rising Gas

This section analyses the influence of rising gas with different flow rate on gas distribution and dome average temperature, and the results as shown in Fig. 7-11 and Fig. 7-12. The inlet area is fixed and the flow rate can be equivalently treated as velocity (v). The initial velocity of the rising gas is quite small, thus, it has little influence on the flow pattern in the dome. When the velocity of the rising gas increases from 0.18m/s to 0.48m/s, meaning that more CO and H_2 enter into the dome. Accordingly, the proportion of effective components in generator gas will increase. In addition, the increasing of velocity results in an obvious spreading effect even if more reducing gas is involved in the oxidation reactions. As shown in Fig. 7-12, with the flow rate of rising gas increases, the dome average temperature increases. For every 880m³/min increase in the rising gas, the average temperature of the dome increases by 24.2K. Therefore, the flow rate of rising gas need to keep in 3357m³/min.

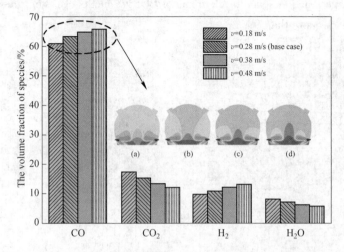

Fig. 7-11　Effect of the velocity of rising gas on composition of the generated gas (Unit: m/s)

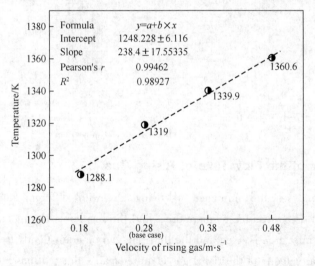

Fig. 7-12　Effect of velocity of the rising gas on the dome average temperature

The composition and temperature of gas along the monitoring line under different flow rate conditions are compared in Fig. 7-12. In reaction zone, the variation tendency of gas composition with flow rate is consistent with the results presented in Fig. 7-11. This is because the reactant gas has a quite low diffusion rate inside the solid particle, which will lead to the heterogeneous reaction being confined to the surface of the solid phase. Hence, at the front of DB, CO and H_2 prefer to combust after contacting O_2 with more heat released. While in the spreading zone, as the flow rate of rising gas increases, the change of gas component and gas temperature show an opposite trend. As Fig. 7-13 shows. the difference becomes more distinct near the central axis of the dome. It can be attributed to the spreading effect of rising gas with a relatively low temperature. In the upper part of the dome, the temperature distribution gradually becomes uniform due to the dilution effect of the gas with different temperatures.

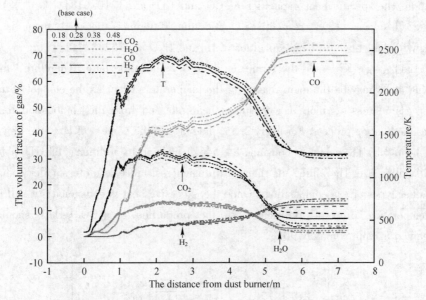

Fig. 7-13　The variation tendency of several variables under different velocity conditions

7.3.4　Effect of the Component of Rising Gas

The depletion of fossil fuels has resulted in an ever-growing challenge for steel manufacturing over the past years. Once the structure of the furnace burden is adjusted in the COREX process, the components of rising gas from the fluidized bed will change first. The influence of different components of rising gas on the thermochemical behaviors in the dome is analyzed. The proportion of ($CO+H_2$) (R) in rising gas is set as 80%, 85%, 90% and 95% and correspondingly decreases the concentration of CO_2 and H_2O. When the flow rate is fixed, the higher the proportion of ($CO+H_2$) is, the more CO and H_2 will enter the dome zone. The combustion reactions occurring in the dome play a role in regulating the temperature and have little impact on the component of the generated gas. It can be seen in Fig. 7-14, when the proportion of ($CO+H_2$) increases by 5%, the proportion of CO and H_2 in the generated gas increases by 2.8%. It can be inferred that more oxygen is consumed by CO and H_2, potentially resulting in the excessive dome temperature and incomplete combustion of the recycling dust.

As shown in Fig. 7-15, the average temperature of the dome drops by 3.4K when the proportion of ($CO+H_2$) in rising gas increases from 90% to 95%. It may be attributed to the excessive CO and H_2 carrying away a large amount of heat. The specific heat capacities of four types of gas in Fig. 7-15 are calculated and compared to verify this assumption. The specific heat capacities of CO and CO_2 are 1210.618J/(kg·K) and 1249.833J/(kg·K), respectively, and the difference between them is very small. While the specific heat capacities of H_2 and H_2O are 15288.411J/(kg·K) and 2369.134J/(kg·K), and the former is about 6.5 times the latter. Therefore, the increasing of H_2 may be the main reason for the temperature drop. As the content of the reducing gas increases, more H_2 consumes heat released from the combustion reaction during the heating process. For every 2.5% increase in the content of H_2, the unit mass of H_2 heated by 1K needs to consume 84.546kJ of heat. In addition, the H_2 molecule has a smaller size. Therefore, its internal diffusion resistance is quite smaller compared with other types of gas compositions. Once the excessive heat is consumed, it will cause the temperature drop of the dome, so the proportion of ($CO + H_2$) should be maintained at approximately 90%.

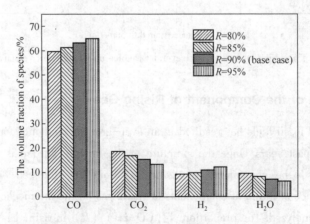

Fig. 7-14 Effect of the component of rising gas on the composition of the generated gas

7.3.5 Effect of the Temperature of Rising Gas

In Fig. 7-16, As the temperature of rising gas increases from 1043K to 1193K, the volume fraction of each gas composition changes insignificantly. With a fixed amount of

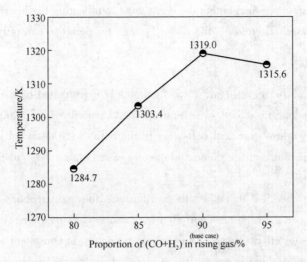

Fig. 7-15 Effect of the component of rising gas on the dome average temperature

oxygen, the temperature increase promotes the diffusion rate of O_2, indicating that more O_2 reacts with the reducing gas. Therefore, the content of CO and H_2 decreases while the content of CO_2 and H_2O increases. The zones with high CO concentration reduce slightly in the vertical profile with the increase of the temperature of rising gas. Therefore, the above analysis once again verifies the conclusion that the composition of gas products is mainly determined by the rising gas.

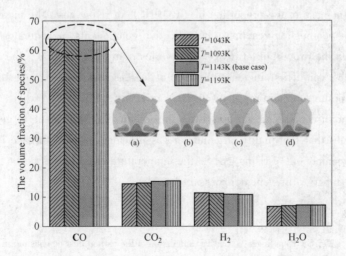

Fig. 7-16 Effect of the temperature of rising gas on composition of the generated gas

The temperature of gas products increases remarkably as the temperature of rising gas increases, and the average value in each stage is 19.432K. A higher temperature of rising gas means that more heat enters the dome zone, indicating the interphase heat transfer will be improved. Therefore, the dome average temperature naturally increases.

7.4 Summary

The combustion of the recycling dust was numerically investigated by a 3-D steady-state mathematical model. The effects of several operating parameters of rising gas on the hydrodynamics and thermochemical behaviors in the dome were analyzed by demonstrating the average temperature of the dome and species distributions. The conclusions are summarized as follows:

According the chemical reaction rate distribution, the main reaction zone can be divided into three part, which are controlled by homogeneous reaction, heterogeneous reaction and diffusion effect in turn. The particle trajectory is consistent with the gas flow pattern. In addition, there is an expansion behavior during the combustion of dust particle. Due to adequate reaction zone and long residence time, the burnout rate of dust particle can reach more than 99.9%.

Increasing the flow rate of rising gas could increase the volume fraction of CO and H_2 in generator gas, while decrease the volume fraction of CO_2 and H_2O. It is worth noting that the temperature in reaction zone and diffusion zone show an opposite variation with the increase of the flow rate of rising gas. However, for every 880m^3/min increase in rising gas, the dome average temperature increases by 24.2K.

With the increase of the proportion of (CO+H_2) in rising gas, the more CO and H_2 participate in combustion reaction, but effective components in generator gas still increases. When the proportion of (CO+H_2) varies from 80% to 95%, the average temperature of the dome first increases and then decreases due to H_2 consume excessive heat. So the optimal proportion of (CO+H_2) in rising gas is 90%.

The temperature of rising gas has little impact on the component of generator gas, which indicates that the quality of generator gas provided to shaft furnace is mainly depend on the component of rising gas. So the temperature of rising gas needs to be keep in a reasonable range to prevent unnecessary heat loss.

References

[1] Berger K, Wei C, Kepplinger W L. CFD simulation of the carbon dust combustion in the COREX® melter gasifier [J]. Steel Research International, 2008, 79: 579~585.
[2] Guo B, Zull P, Rogers H, et al. Three-dimensional simulation of flow and combustion for pulverized coal injection [J]. ISIJ International, 2005, 45: 1272~1281.
[3] Liu Y, Shen Y. Three-dimensional modelling of charcoal combustion in an industrial scale blast

furnace [J]. Fuel, 2019, 258: 116088.
[4] Du K, Wu S, Zhang Z, et al. Analysis on inherent characteristics and behavior of recycling dust in freeboard of COREX melter-gasifier [J]. ISIJ International, 2014, 54: 2737~2745.
[5] Wang Q, Zhang J, Wang G. Simulation study of dust combustion in COREX gasifier dome [J]. Ironmaking Steelmaking, 2020, 47: 344~350.
[6] Xu R, Zhang J, Wang W, et al. Factors influencing gas generation behaviours of lump coal used in COREX gasifier [J]. High Temperature Materials and Processes, 2019, 38: 30~41.
[7] Qu J, Wu K, Wang C, et al. Effect of pulverized coal injection on the theoretical flame temperature before tuyere in COREX melter-gasifier [J]. Iron and Steel, 2020, 47 (6): 19~22.
[8] Lu L, Gao X, Dietiker J, et al. MFiX based multi-scale CFD simulations of biomass fast pyrolysis: a review [J]. Chemical Engineering Science, 2020, 248: 117131.
[9] Smith T F, Shen Z F, Friedman J N. Evaluation of coefficients for the weighted sum of gray gases model [J]. Asme Journal of Heat Transfer, 1982, 104 (4): 602~608.
[10] Zhou C, Wang Y, Jin Q, et al. Mechanism analysis on the pulverized coal combustion flame stability and NO_x emission in a swirl burner with deep air staging [J]. Journal of the Energy Institute, 2018, 92 (2): 298~310.
[11] Schiller L, Naumann A. A drag coefficient correlation [J]. VDI Z, 1935, 77: 318~320.
[12] Abdelghany A, Fan D, Sohn H. Novel flash ironmaking technology based on iron ore concentrate and partial combustion of natural gas: a CFD study [J]. Metallurgical and Materials Transactions B, 2020, 51: 2046~2056.
[13] Huang W, Pourkashanian M, Ma L, et al. Effect of geometric parameters on the drag of the cavity flameholder based on the variance analysis method [J]. Aerospace Science & Technology, 2012, 21 (1): 24~30.
[14] Verma K A, Pandey K M, Ray M, et al. Effect of transverse fuel injection system on combustion efficiency in scramjet combustor [J]. Energy, 2021, 218 (1): 119511.
[15] Zaitri M, Bouchetara M, Bouziane A, et al. Effect of CH_4-H_2 mixture on the combustion characteristics in a stabilized swirl burner [J]. International Journal of Energy Research, 2021, 46 (3): 2923~2933.

8 Numerical Simulation of Pulverized Coal Injection in the Dome Zone of COREX Melter Gasifier

In COREX melter gasifier (MG), the secondary injection systems are installed in the dome zone to recycle dust and oxygen. The oxygen burner in the dome zone can be changed into an oxygen-coal burner for pulverized coal injection (PCI) by referring to the coal gasification technology. The pulverized coal and oxygen can be injected into the dome zone so that gasification can occur when partial oxidant is supplied. On the premise of meeting the smelting heat demand of the packed bed and hearth zone, the PCI in the dome zone can partly undertake the function of generation of reducing gas from lump coal in the lower part of the MG, and further greatly improve the quality of reducing gas and promote the reduction process in the shaft furnace. On the other hand, injecting low-cost powdered coal can reduce the production cost and control the total fuel ratio in COREX. Hence, it is necessary to study the PCI in the dome zone of MG for the optimization and development of the COREX process.

The COREX MG is a 'black box' reactor, and it is difficult to directly observe and measure the inner characteristics under high temperature and high pressure conditions. The numerical simulation method that couple heat, mass, and momentum balance is regarded as an alternative for investigating and optimizing the PCI in the dome zone of COREX MG. Computational fluid dynamics (CFD) has been widely used to study coal gasification in gasifier, coal combustion in the raceway of blast furnace. For the reaction of the carbonic particles in the dome zone of the COREX MG, the combustion behavior of recycling carbon dust in front of the dust burner were studied. These works, especially the CFD model, provide an effective reference for the exploration of PCI in the dome of COREX MG. However, the different properties between recycling dust and pulverized coal, and the corresponding operation makes it impossible for the latter to draw experience directly from recycling dust injection. The recent practice of PCI in the dome zone of COREX MG showed that the amount of generated reducing gas increased by 15%, and the metallization ratio in the shaft furnace increased by 15% ~ 20%. However, detailed information about the PCI in the dome zone of COREX MG is still rare, especially the quantitative comparisons of multiphase reaction flow and reducing gas composition with and without PCI have not been reported.

In this chapter, a three-dimensional mathematical model is developed to study the PCI in the dome zone of COREX MG. The transport phenomenon in the dome zone with and without PCI are compared with a particular focus on the velocity, temperature, gas composition, and particle behavior. The findings of this work lay a foundation for the further research of COREX MG operation of PCI in the dome zone, and provide a theoretical basis for guiding the actual industrial production and application.

8.1 Mathematical Model

8.1.1 Governing Equations

In the simulation of the PCI in the dome zone of COREX MG, there exists a complex flow, heat, and mass transfer phenomenon, which follows four governing equations including mass, momentum, energy, and species conservation. The gas phase is treated as a continuous phase with the assumption of an incompressible ideal gas. The governing equations can be expressed by the following general partial differential formulas.

$$\frac{\partial(\rho\phi)}{\partial t} + \text{div}(\rho U \phi) = \text{div}(\Gamma_\phi \text{grad}\phi) + S_\phi \quad (8-1)$$

From the left, each term represents the unsteady term, convection term, diffusion term, and source term in turn, where the source term is the term that cannot be included in the first three terms. In this study, a 3-D steady mathematical model is developed, so the steady term is not considered.

The gases are sprayed into the dome zone at a rapid velocity, and turbulence is fully developed. To predict the turbulent mixing and diffusion trajectory of particle phase in high-speed flow more accurately and effectively, the Realizable $k-\varepsilon$ double equation is used to close the steady-state Reynolds time-averaged Navier-Stokes equations. The transport equations of turbulent kinetic energy k and turbulent dissipation rate ε are shown as follows.

$$\frac{\partial(\rho k u_i)}{\partial x_i} = \frac{\partial}{\partial x_j}\left[\left(\mu + \frac{\mu_t}{\sigma_k}\right)\frac{\partial k}{\partial x_j}\right] + G_k + G_b - \rho\varepsilon \quad (8-2)$$

$$\frac{\partial(\rho \varepsilon u_i)}{\partial x_i} = \frac{\partial}{\partial x_j}\left[\left(\mu + \frac{\mu_t}{\sigma_\varepsilon}\right)\frac{\partial \varepsilon}{\partial x_j}\right] + \rho C_1 S_\varepsilon - \rho C_2 \frac{\varepsilon^2}{k + \sqrt{v\varepsilon}} + C_{1\sigma}\frac{\varepsilon}{k}C_{3\sigma}G_b \quad (8-3)$$

where G_k and G_b are the turbulent kinetic energy term generated by laminar velocity gradient and buoyancy respectively. $C_1 = 1.44$, $C_2 = 1.92$, $\sigma_k = 1.0$, $\sigma_\varepsilon = 1.2$, σ_k and σ_ε are the turbulence Prandtl numbers of the k equation and ε equation.

The recycling dust and pulverized coal are regarded as discrete particle phases. Due to the volume loading of particle phase in the dome zone being less than 10%, the Lagrangian method is adapted. Once particles enter into the dome zone through the burner, they will be rapidly diluted by other gas phase components. Therefore, the interaction between the particles is not considered. The Random Walk Model (RWM) is used to describe the trajectory of particles and the motion of particles follows Newton's second law.

$$\frac{du_p}{dt} = F_D(u - u_p) + \frac{g_x(\rho_p - \rho)}{\rho_p} + F_x \tag{8-4}$$

where $F_D(u-u_p)$ is the drag force of particle in per unit mass, u_p is particle velocity, u is gas velocity, ρ_p and ρ are the density of particle and gas respectively, F_x is the sum of other forces on the particle. F_D can be calculated by Eq. (8-5), where C_D is the coefficient of drag force.

$$F_D = \frac{18\mu}{\rho_p d_p^2} \frac{C_D Re}{24} \tag{8-5}$$

There are three main methods of heat transfer to decide the temperature change of particles, including convective heat transfer, sensible heat caused by mass transfer, and radiation heat transfer. The radiation heat transfer is solved by the P-1 radiation model, in which the radiation coefficient is evaluated by the gray gas weighted average model.

8.1.2 Chemical Reaction Model

The chemical reactions that occur in the dome zone are a multi-stage coupling complex process. This process can be divided into devolatilization, homogeneous reactions, and heterogeneous reactions. Hence, careful consideration was given to the selection of these reaction sub-models.

8.1.2.1 Devolatilization Model

The evaporation process of water is not considered because both dust and pulverized coal have been preheated to above 100℃ before entering the dome. It is worth noting that the combustible carbon in the dust particle originates from the unburned coal and coke. Therefore, in addition to pulverized coal, dust particles need to consider the release of volatile in the heating process. The two-competing reactions model assumes that there is a pair of parallel first order irreversible reactions in the process of devolatilization, which has been applied in many related simulations.

$$\text{Dust+Raw coal} \begin{array}{c} \xrightarrow{k_1} a_1 VM_1 + (1-a_1)C_1 \quad \text{Low temperature} \\ \xrightarrow{k_2} a_2 VM_2 + (1-a_2)C_2 \quad \text{High temperature} \end{array} \tag{8-6}$$

where reaction rate constants k_1 and k_2 are expressed by the Arrhenius formula as shown in Eq. (8-7).

$$k = A\exp(-E/T) \quad (8-7)$$

Among them, A is the pre-exponential factor and E is the activation energy. At low temperatures, A and E are $3.7 \times 10^5 \mathrm{s}^{-1}$ and 1.5×10^8 J/kmol respectively. While at high temperature, A and E are 1.46×10^{13} s^{-1} and 2.51×10^8 J/kmol respectively. Volatile is an organic compound consisting of various elements and it can be expressed as $C_\alpha H_\beta O_\gamma N_\theta$. When the temperature condition is satisfied, the decomposition process of volatile is assumed to occur instantaneously. The mixtures of CO, CO_2, CH_4, H_2, and N_2 are regarded as the decomposition products of volatile. Based on the elemental analysis of dust and coal, the stoichiometric coefficient of each component can be estimated, and this process can be described as Eq. (8-8).

$$C_\alpha H_\beta O_\gamma N_\delta \rightarrow m_1 CO + m_2 CO_2 + m_3 CH_4 + m_4 H_2 + m_5 N_2 \left(\sum_{i=1}^{5} m_i = 1\right) \quad (8-8)$$

8.1.2.2 Heterogeneous Reactions

After the release of the volatile, the residual char will participate in the heterogeneous reactions, including the reactions of $C(s)-O_2$, $C(s)-CO_2$, and $C(s)-H_2O$. The multiple surfaces reactions model assumes that the particle surface species can be depleted or produced by the stoichiometry of the particle surface reaction. The particle surface species of dust and pulverized coal are regarded as carbons, participating in oxidation and gasification reactions of residual char. The Arrhenius formula is used to calculate the kinetic rate of reaction r.

$$k_r = A_r T_p^\beta e^{-(E_r/T_p)} \quad (8-9)$$

The apparent order of reaction r is 1, so the rate of particle surface carbon depletion $R_{j,r}$ is given by:

$$R_{j,r} = A_p \eta_r Y_j P_n \frac{k_r D_{0,r}}{D_{0,r} + k_r} \quad (8-10)$$

where A_p and η_r are particle surface area (m^2) and effectiveness factor, respectively. Y_j is the mass fraction of surface j in the particle. P_n represents the bulk partial pressure of the gas phase species n (Pa). The diffusion rate coefficient $D_{0,r}$, for the reaction r is defined as Eq. (8-11):

$$D_{0,r} = C_{1,r} \frac{[(T_p + T_\infty)/2]^{0.75}}{d_p} \quad (8-11)$$

$C_{1,r}$ indicates the molar concentration of species j in reaction r. The related kinetic parameters are summarized in Table 8-1.

Table 8-1 Kinetic parameters of chemical reactions

Heterogeneous Reactions	A_r/ kg · m² · Pa^{-N} · s^{-1}	β	E_r/J · kmol^{-1}
$C(s) + 0.5O_2 \rightarrow CO$	1.36×10^6	0.68	1.30×10^8
$C(s) + CO_2 \rightarrow 2CO$	6.78×10^4	0.73	1.63×10^8
$C(s) + H_2O \rightarrow CO + H_2$	8.55×10^4	0.84	1.40×10^8

Except for the reactions mentioned above, the oxidation reactions of alkalis and reduction reactions of iron oxides are not considered in this simulation. This is because the enthalpies involved in these reactions have little effect on the overall temperature of the dome compared with the combustion or gasification reactions.

8.2 Simulation Conditions

In this study, the geometric model is constructed regarding COREX-3000 industrial data, as shown in Fig. 8-1. The geometric model is roughly a hemispherical space, with four generator gas outlets in the upper part, which are set as pressure outlets. The upper boundary of the packed bed zone is set as the bottom inlet of the mathematical model, where the rising reducing gas enters the dome zone. There is a row of circumferential burners arranged in the shrink part of the dome, and the angle between the tip of the lance and the horizontal line is 8°. For the case of the recycling dust injection in the dome zone (regarded as case-1 in this work), six oxygen burners and four dust burners at the lower part of the geometry. Their distributions and positions in the present study are consistent with that in the literature. Industrial pure oxygen (99.5% O_2) is introduced by cylindrical hollow pipes in OB. Meanwhile, the DB is equipped with a co-axial lance, of which the oxygen is transported in an external tube and the dust is injected in the inner tube with a carrier gas (100% N_2). This structure could promote the formation of a shear layer of oxygen flow, which is conducive to improving the burnout rate of dust particles. For the coal injection in the dome zone (regarded as case-2), four OB (located at the 22.5°, 157.5°, 202.5°, 337.5° positions along the circumferential direction) are converted to oxygen coal burners (OCB), and the DB maintains the original structure and position, as shown in Fig. 8-2. The OCB is also a co-axial lance, but unlike the DB, the oxygen is blown in through the inner tube and the coal particles are injected from the external tube. This structure can prevent excessive oxygen combusting with the rising reducing gas and enable more pulverized coal particles to gasify under insufficient oxidant conditions. All entrances in the model are treated as velocity inlets. The furnace wall is simplified as the non-slip shear condition, and there is no existing mass penetration. In practical production, the burden from the upper shaft furnace charging into the dome zone will take away the heat, and the charging lump coal in the top of the dome zone also will consume the heat. In this work, these heat are calculated by a static model considering the mass and heat balances, and then the value is added as an energy source term in the governing equation.

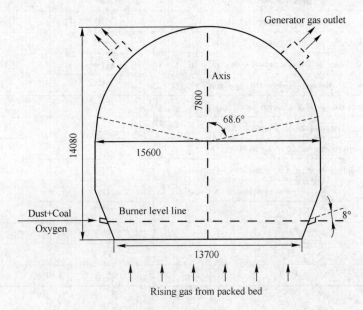

Fig. 8-1 Schematic diagram of the mathematical model (Unit: mm)

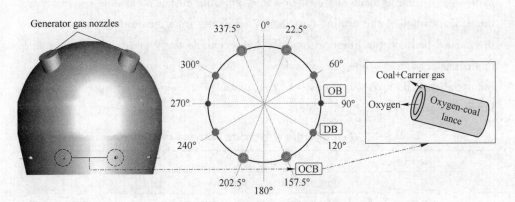

Fig. 8-2 Schematic diagram of the injection systems installed in the dome zone

The operation parameters of MG under the condition of recycling dust injection (case-1) and PCI (case-2) are shown in Table 8-2. The oxygen supply in both DB and OB decreases in case-2, and more oxygen is supplied through OCB, which could form the temperature and pressure conditions for satisfying the gasification of coal. In addition, the composition of rising gas coming from the packed bed also changes. All these operating parameters are obtained from the plant operation.

Table 8-2　Operation parameters of MG under the condition of case-1 and case-2

Operating Parameters	Case-1	Case-2
Generator gas pressure/kPa	350	350
Oxygen supplied by OB (standard state)/$m^3 \cdot h^{-1}$	3975	2478
Oxygen supplied by DB (standard state)/$m^3 \cdot h^{-1}$	11804	8829
Recycling dust supplied by DB/$kg \cdot min^{-1}$	390	370
Flow rate of rising gas (standard state)/$m^3 \cdot min^{-1}$	2477	2099
Temperature of rising gas/K	1143	1143
Pulverized coal supplied by OCB/$kg \cdot t^{-1}$	—	150
Oxygen supplied by OCB (standard state)/$m^3 \cdot h^{-1}$	—	18548
Gas composition of rising gas (volume fraction)/%		
CO	71.1	71.2
CO_2	4.8	11.0
H_2	18.9	12.2
H_2O	2.6	3.1

The properties of recycling dust and coal particles are shown in Table 8-3. The proximate analysis and ultimate analysis show that recycling dust is a valuable secondary material. The carbon in the cycling dust mainly originates from unconsumed fine coke from the packed bed and the fine coal from the lump coal. So the volatiles of recycling dust, approximately 4.6%, must be considered in this work. The properties of coal refer to industrial data. Both recycling dust and coal particles are assumed a uniform size distribution is uniform as shown in Table 8-3.

Table 8-3　The properties of recycling dust and coal (dry basis)

Properties	Recycling dust	Coal
Proximate analysis (mass fraction)/%		
Ash	68.22	8.85
Volatile	4.60	33.41
Fixed carbon	27.18	57.74
Ultimate analysis (mass fraction)/%		
C	89.68	83.30

Continued Table 8-3

Properties	Recycling dust	Coal
H	1.13	4.87
O	8.59	10.86
N	0.60	0.97
Colorific value/MJ · kg^{-1}	25.10	30.14
Density/kg · m^{-3}	3800	1400
Particle size/μm	28.8	80

Although this model is developed based on industrial data, some inevitable simplifications are made to improve the efficiency, including:

(1) The composition and temperature of the rising gas are regarded as a uniform diffusion when it enters into the dome zone.

(2) No adherence or agglomeration occurs during the motion of particles.

(3) Dust particles-wall collisions do not influence particle heat and mass transfer.

(4) CH_4 is produced by the decomposition of volatile in the pulverized coal.

The simulation work is completed based on Fluent 14.5. The governing equations, turbulent kinetic energy, turbulent dissipation, and other equations are solved by the finite volume method. The SIMPLE semi-implicit algorithm is used to calculate the pressure-velocity coupled equations. The convergence standard requires the residual curves of each index to be less than 1×10^{-5}, especially the residual curves of energy and P1 are less than 1×10^{-6}. At the same time, monitoring the variation trend of H_2 concentration at the outlet and the dome average temperature to judge the convergence of simulation results. When the average relative error of variables before and after iteration is less than 0.1%, the simulation result is regarded as convergence.

8.3 Results and Discussion

8.3.1 Model Validation

The mathematical model was validated by comparing the industrial test data and simulation results from the perspectives of composition of generator gas at the outlet and dome average temperature, as shown in Table 8-4. Although the model is developed based on industrial conditions, some extremely complex behaviors are simplified, so there are some small differences between simulation data and industrial data. But in general, the simulated data has reached the requirement of industrial production. For example, the relative error of the dome average temperature is only 0.047%, and the proportion of

CO and H_2 reaches 83.3%, which could provide sufficient reductants for the reduction of ore in the shaft furnace. Therefore, the developed CFD model in this study is reliable and can be used to predict the transport phenomenon in the dome zone of COREX MG with PCI.

Table 8-4 Comparison between industrial data and simulation data

Parameters	Industrial data	Simulation data
Dome average temperature/K	1375	1374.35
CO/%	66.47	67.48
CO_2/%	7.72	8.58
H_2/%	16.25	16.11
N_2/%	6.08	5.80
Other gases/%	3.47	2.306
Effective components/%	82.72	83.28

8.3.2 Effect of PCI in Dome Zone on the Performance of COREX MG

In this section, a detailed comparison of the velocity field, temperature field, gas composition distribution, and particle behavior in the dome zone of COREX MG with single recycling dust injection (case-1) and dust-pulverized coal co-injection (case-2) were discussed.

8.3.2.1 Velocity Field

Fig. 8-3 shows the velocity field in the dome zone of case-1 and case-2. Both the gas velocities show a uniformity and symmetry distribution. It can be seen from Fig. 8-3(a) and (b), in case-1, the positions of the burners are oppositely distributed, so after the high-speed fluid enters the dome zone through the burners, the gas collides at the center zone and causes a radial diffusion flow with a velocity about 7m/s. There is no obvious velocity gradient near the wall zone, which is easy to form circulation and increases the residence time of particles, which is conducive to the complete combustion or gasification of dust particles. In case-2 with PCI in the dome zone, the pulverized coal particles are injected through the oxygen coal burner, so the oxygen supply at the oxygen-coal burner is about 2.1 times that of the dust burner, thus the velocity in front of the oxygen coal burner is much larger as shown in Fig. 8-3(c). In addition, the increase of the zone with high velocity as shown in Fig. 8-3(d), will lead to the obvious expansion of the impact area between gas phases, which may affect the complex chemical reaction zone in the lower part of the dome.

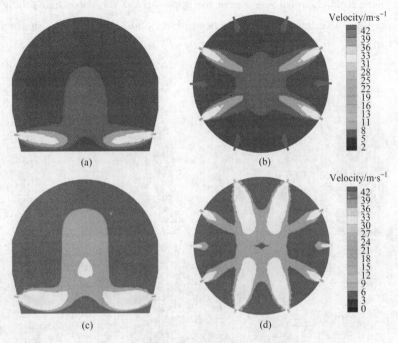

Fig. 8-3　Velocity field in the dome zone
(a) Case-1 DB vertical section; (b) Case-1 horizontal section;
(c) Case-2 OCB vertical section; (d) Case-2 horizontal section

8.3.2.2　Temperature Field

The temperature distribution in case-1 with single recycle dust injection is shown in Fig. 8-4. The oxygen with high speed is blasted into the dome zone through the outer pipe of the DB, and it will entrain with the surrounding dust particles and reducing gas, to burn rapidly and form a flame peak, with a maximum temperature of 2400 K. It can be found from Fig. 8-4(a) that the contour of the lower half of the flame peak is significantly smaller than that of the upper half. The main reason is that the momentum direction of the lower half of the flame peak is opposite to the buoyancy direction caused by the rising gas from the packed bed of COREX MG, which hinders the development of the lower half of the flame peak. In addition, due to the release process of volatile in dust particles, there will be a low-temperature area with volatile enrichment in the interior of the flame peak, so the flame summit will gradually develop in a ring shape. This phenomenon can be verified by the layered contour diagrams in front of DB as shown in

Fig. 8-4(b). The oxygen blasted through OB is used to adjust the dome temperature, and its oxygen flow is low, so the temperature contour in front of the oxygen burner is smaller. There is no chemical reaction in the upper area of the dome zone, only affected by the plug flow. The temperature will be uniform and stable finally, maintaining at about 1330K.

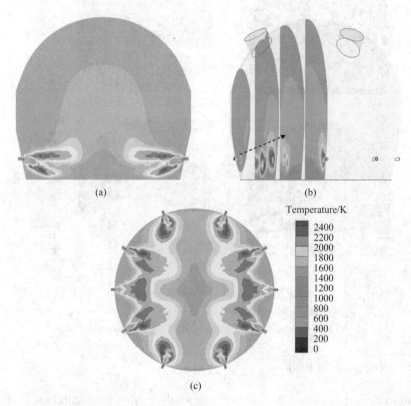

Fig. 8-4 Temperature distribution in dome zone under case-1 condition
(a) DB vertical section; (b) Longitudinal sections in front of DB; (c) Horizontal section

Fig. 8-5 shows the temperature distribution in the dome zone with PCI. It can be seen the temperature distribution in the dome zone changes significantly due to the redistribution of oxygen flow. From Fig. 8-5(a), it can be seen the high temperature zone in front of DB in case-2 is somewhat smaller than that in case-1. This is due to the decrease of oxygen volume at the dust burner in case-2, while the flame structure is basically consistent with that in case-1. For the OCB, the oxygen is supplied through the inner pipe and a columnar high-temperature zone (about 2600K) is formed, as shown in the red zone in Fig. 8-5(b). In this high temperature zone, the temperature first increases and then decreases. The main reason is that the pulverized coal particles and oxygen are firstly burned, and a large amount of heat is released, then the endothermic reactions such as $C(s)-CO_2$, $C(s)-H_2O$, $CO-H_2O$, and CH_4-H_2O dominate. The temperature distribution is closely related to the chemical reaction. The transfer of the high temperature zone indicates that after the PCI in the dome zone, the main reaction zone is transferred to the front of the OCB. Finally, the temperature of generator gas is stable at about 1370K. Compared with case-1, the dome average temperature in case-2 increases by about 40K.

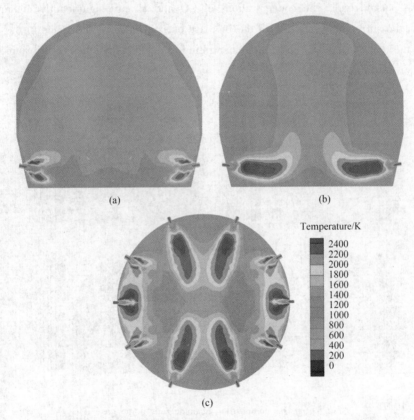

Fig. 8-5 Temperature distribution in dome zone under case-2 condition
(a) DB vertical section; (b) OCB vertical section; (c) Horizontal section

8.3.2.3 Gas Composition

The composition of generator gas at the outlet in MG is directly related to the quality of reduction gas used in the upper COREX shaft furnace. Fig. 8-6 shows the gas distribution in case-1 on the vertical section of DB. It can be seen that the volume fractions of CO and H_2 are opposite to that of CO_2 and H_2O respectively. Where the concentrations of CO and H_2 are high, the concentrations of CO_2 and H_2O are low. In the front of the DB, the oxygen concentration is sufficient, and the dust and the reducing gas rising from the packed bed have a fierce combustion reaction with O_2, so the concentration of CO_2 and H_2O is the highest. At the end of the gas stream, most of the oxygen is consumed. While the secondary reaction of carbon residue ($C(s)-CO_2$ and $C(s)-H_2O$) is fully developed, and the concentration of CO increases rapidly, up to 85%. In the central zone of the dome, due to the limited combustion reaction and the diffusion effect of rising gas from the packed bed, the concentrations of CO and H_2 are significantly higher than those in the surrounding area. With the increase of axial height, the space in the upper part of the dome expands, and the concentration of each gas reaches equilibrium under the action of diffusion flow.

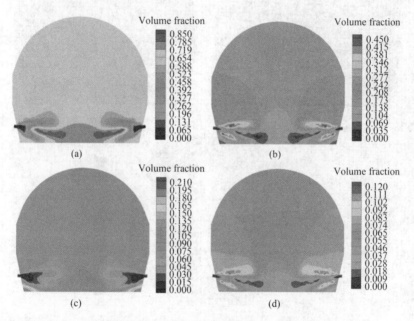

Fig. 8-6 Gas composition in dome zone under case-1
(a) CO; (b) CO_2; (c) H_2; (d) H_2O

The volume fraction of reducing gas (CO and H_2) in case-2 is shown in Fig. 8-7. It is obvious that concentration distribution of CO_2 and H_2O are opposite to that of CO and H_2 respectively, so the distribution of concentration of CO and H_2, which are more important to the reduction processes in COREX, are discussed. As can be seen in Fig. 8-7 (a), due to the decrease of gas volume, the CO enrichment area in front of DB is closer to the furnace wall compared with that in case-1. In addition, affected by the gasification reaction near the OCB, there appears a region with a high concentration of H_2 at the center of the DB vertical section, shown in Fig. 8-7(b). On the vertical section of the OCB, an H_2 enrichment region similar to that in Fig. 8-7(b) appears. This is because in the latter half of the flow, the oxygen concentration is low and multiple gasification reactions are fully developed, resulting in the generation of a large amount of H_2.

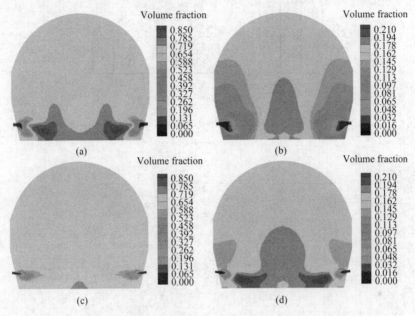

Fig. 8-7 Gas composition in dome zone under case-2
(a) CO on DB vertical section; (b) H_2 on DB vertical section;
(c) CO on OCB vertical section; (d) H_2 on OCB vertical section

Fig. 8-8 shows the comparison the proportion of reducing gas in generator gas at the outlet of the dome zone. It can be seen that with a total of 150kg/t coal injected into the dome zone, the proportion of CO_2 in the reducing gas decreases by 7.8%, and the volume fraction of H_2 increases by 7.4%. The volume fraction of CO does not increase greatly. In total, the proportion of the effective component in generator gas increases by 10.9%. In addition to meeting the demand of the upper reduction process in the COREX shaft furnace, this hydrogen-rich reducing gas can also be blasted into a blast furnace for reducing the coke ratio.

To further explore the effect of PCI in the dome zone on the concentration distribution of components, the variation tendency of concentration of CO and H_2 along the central axis of case-1 and case-2 are compared, as shown in Fig. 8-9. It can be seen that due to the interaction of complex chemical reactions, the volume fraction of CO and H_2

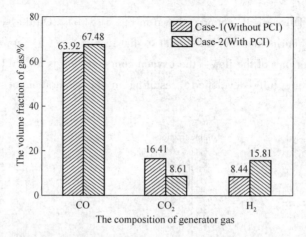

Fig. 8-8 Volume fraction of reducing gas at the outlet of the dome under case-1 and case-2

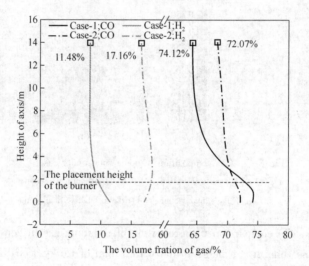

Fig. 8-9 Variation of concentration of reducing gas along with the height on the central axis of case-1 and case-2

change significantly at the placement height of burner. In case-1, dust combustion is the dominant process, thus some rising reducing gas from the packed bed is burned to adjust the dome temperature. Therefore, the volume fraction of CO and H_2 decreases sharply at the placement height of burner. For case-2, the PCI in the dome zone is mainly to realize the coal gasification. It can be found that the volume fraction of H_2 at the placement height of the burner increases, while the volume fraction of CO decreases. The latter is consumed by the water-gas shift reaction. In addition, there are more complex gas components and reaction mechanisms in the dome after PCI, so the concentration fluctuation of CO and H_2 is more obvious above the placement height of the burner. The generator gas outlet is located in the edge area of the dome as shown in Fig. 8-1, which is more vulnerable to the diffusion of reaction products than the central area of the dome. Therefore, there are some differences between the gas concentration at the top center and the data at the gas outlet.

8.3.2.4 Particle Behavior

The trajectory and transformation behaviors of coal particle are very important for the stable operation of the dome zone. The burnout rate is an important index to evaluate conversion extent of carbon in the particle, which has been accepted by many scholars. It can be expressed as Eq. (8-12).

$$\text{Burnout} = \left(1 - \frac{m_{a,0}}{m_a}\right) \Big/ (1 - m_{a,0}) \tag{8-12}$$

where $m_{a,0}$ is the ash content of the original dust and m_a is the ash content of the burning residual. Fig. 8-10 shows the residence time and burnout rate of particles in case-1 and case-2. It can be seen that the trajectories of particles are consistent with the velocity field, indicating that the motion of particles are mainly affected by the gas flow. In case-1, the longest residence time of dust particle in the dome zone can reach about 70s, while in case-2, the longest residence time of particles increased by about 10s. Although the gas velocity increases in the dome zone, the aggravating circulation flow will lengthen the residence time of particles. In addition, the simultaneous injection of pulverized coal and dust will increase the number of particles, resulting in more complex movement behavior in the dome zone.

From Fig. 8-10(b), it can be found that the outer edge of dust particle jet and surrounding oxygen form entrained flow, which can promote the complete combustion of dust particle. While in case-2, the inner edge of coal particle stream that is fully in contact with oxygen shows a better combustion behavior. Although there are some residual carbon in both particles, the burnout rate at the end of jet is still nearly 100%. This also verifies that after oxygen depletion, the remaining carbon in the particles do participates in the gasification reaction. In addition, the long residence time of dust and pulverized coal particles and the sufficient reaction space in the dome can also improve the conversion rate if both particles, which can finally reach more than 99%.

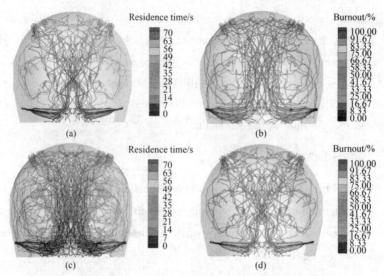

Fig. 8-10 Residence time and burnout rate distribution of particles
(a) Residence time in case-1; (b) Burnout rate in case-1;
(c) Residence time; (d) Burnout rate in case-2

8.4 Conclusions

With the intent to gain a better understanding of the performance of a COREX melter gasifier for pulverized coal injection, a three-dimensional mathematical model is developed to study the transport phenomenon in the dome zone of COREX MG with and without PCI. The comparison of the gas flow field, temperature distribution, gas composition, and particle trajectories is discussed.

The following conclusions could be drawn from the present study:

(1) The gas flow field in the dome can be divided into a high-speed jet zone, impinging flow zone, diffusion flow zone, and circulation flow zone. In the case of PCI, the increase of high-speed jet beam leads to the expansion of impinging flow area.

(2) In the case of PCI in the dome zone, due to the redistribution of oxygen flow, the high-temperature zone in the dome zone is transferred, and the furnace top temperature increases by about 40K.

(3) When 150kg/t coal is injected into the dome zone, the proportion of CO_2 in the reducing gas decreases by 7.8%, while the proportion of H_2 increases by 7.4%. The total effective volume fraction of reducing gas in the outlet of MG can reach 83.3%.

(4) After PCI in the dome zone, the residence time of particles increases by about 10s, which is caused by the aggravating circulation and the complexity of particle behavior. The burnout rate of dust and pulverized coal particles can reach more than 99%.

References

[1] Tian B, Ji S. Improvement and practice of coal injection on the arch roof of the OY furnace in Bayi Steel [J]. XinJiang Iron Steel. 2018, 4: 37~39.

[2] Xu J, Liang Q, Chen R, et al. Study of smelting reduction iron coupled with pulverized coal gasification to produce high-concentration syngas [J]. Scientia Sinica Technologica, 2021, 51: 195~206.

9 Mathematical Study the Top Gas Recycling into COREX Melter Gasifier

Top gas recycling (TRG) technology is an effective means to reduce fuel consumption and CO_2 emission in the ironmaking field. In COREX process, COREX export gas is produced in large quantities as a by-product. COREX export gas has a relatively high lower heating value (LHV), so the gas may be a highly valuable fuel for metallurgical combined heat and power (CHP) plant, and it can be used as a substitute for natural gas. Because the COREX process is a typical nitrogen-free process, COREX export gas can be utilized as a reducing agent in the direct reduction process (e. g. MIDREX). In addition, COREX export gas can also be used as raw material for the production of chemicals. Lampert et al. presented the possibility of integration of the COREX process, blast furnace, CO_2 removal installation and metallurgical CHP plant. The COREX export gas is proposed to be injected as reducing agent into the thermal reserve zone of the blast-furnace process. And the results show that this operation can decrease the consumption of coke. From these studies, it can be seen that the top gas recycling technology is an effective method to reduce the carbon emissions of the ironmaking process, and the large amount of gas produced by the COREX process is also widely used. However, there are few researches on the COREX process combined with the top gas recycling technology.

In this study, a mathematical model of COREX process was developed based on mass and heat balances. Based on this model, according to the actual production data of plant, the operation parameters of COREX process were calculated. The influence of three different types of top gas recycling, such as the recycle gas without CO_2 removal (RG-1), the recycle gas after CO_2 removal (RG-2) and the heated recycle gas after CO_2 removal (RG-3), on the operation parameters and CO_2 emissions of COREX process is investigated.

9.1 Mathematical Modelling

9.1.1 Description

The COREX process combined with top gas recycling technology is showed schematically in Fig. 9-1. In addition to the reduction shaft and melter-gasifier, the CO_2 removal equipment is also considered in the whole process.

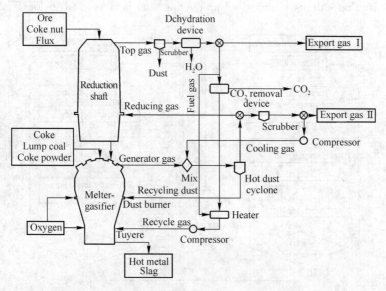

Fig. 9-1 The schematic diagram of investigated COREX process with top gas recycling

Based on the actual plant data of COREX process, the material flow and energy flow of COREX process are calculated by the model that developed by mass and heat balances. According to the calculation results, the current operation condition of COREX process is understood, and the theoretical basis is provided for improving operation parameters. The material balance part is the basis of this model, and the scale of the model is based on producing 1t hot metal. First, the actual productions data of sinter, pellets, lump ores, fluxes, coke, pulverized coal, semi–coke, coke powder, hot metal, DRI, dust, oxygen and other materials are input into the model. Second, the consumptions of raw materials are calculated based on the production of 1t hot metal. Furthermore, based on the balance of Fe, C, H, O, N and other elements, the material flow in the reduction shaft and melter-gasifier is calculated, and the material balance of the reduction shaft and melter-gasifier is achieved by iterative calculation. Then the material balance part of COREX process is obtained, which provides parameters for the study of heat balance. The calculation of heat balance is based on material balance. The reduction shaft and melter-gasifier are taken as the research objects, and the heat balance of the two reactors is investigated respectively. The corresponding heat balance model is established by calculating the heat income items, such as the amount of heat brought into the furnace by the raw materials, the heat emitted by the carbon combustion etc., and the heat output items, such as the amount of heat taken away by the gas, the heat consumed by the direct reduction reaction etc. The concept of the base case that only uses the actual production data is illustrated in Fig. 9-2.

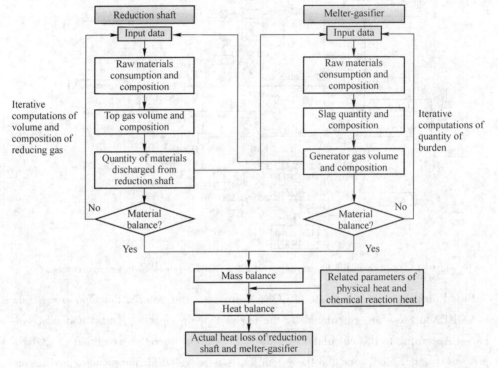

Fig. 9-2　The concept of the base case that only uses the actual plant data

In the COREX process with top gas recycling, the combustion and gasification of carbonaceous fuels, not only generate gas and heat, but also create a lot of space for the continuous falling of burden, which directly affects the falling rate of burden, and then affects the smelting rate of the COREX furnace. When the recycle gas is injected in the tuyere raceway, the amount of oxidizing gas in the tuyere raceway will be increased, and the consumption of carbonaceous fuel will also be increased, thus affecting the smelting rate. In order to keep the smelting rate stable, it is necessary to adjust the amount of oxygen injected at the tuyere to ensure the constant fuel rate in the tuyere raceway. The concept of the cases with top gas recycling is illustrated in Fig. 9-3.

9.1.2　Establishment of the Mathematical Model

The actual production data is a necessary condition for establishing the model. The actual chemical compositions of raw materials and fuels are shown in Tables 9-1 ~ 9-3. In the

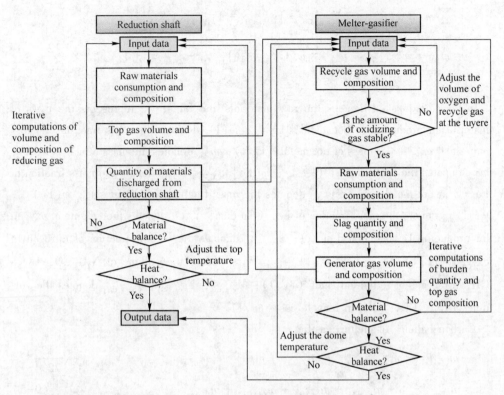

Fig. 9-3 The concept of the cases with top gas recycling

modeling process, the whole system should follow mass and heat balances. The main balance equations of this mathematical model are as follows:

Fe balance of COREX process:

$$\sum_i m_i \times w(\text{Fe})_i = m_{\text{HM}} \times w[\text{Fe}]_{\text{HM}} + m_{\text{slag}} \times w(\text{FeO})_{\text{slag}} \times \frac{56}{72} + m_{\text{dust}} \times w(\text{Fe})_{\text{dust}}$$

(9-1)

Hydrogen balance of COREX process:

$$\sum_i m_i \times w(\text{H})_i + \sum_i m_i \times w(\text{H}_2\text{O})_i \times \frac{2}{18}$$
$$= \frac{2 \times V_{\text{tg, H}_2} + 2 \times V_{\text{tg, H}_2\text{O}} + 4 \times V_{\text{tg, CH}_4}}{22.4}$$

(9-2)

Nitrogen balance of COREX process:

$$\sum_i m_i \times w(\text{N})_i \times \frac{22.4}{28} + V_{\text{oxygen}} \times x_{\text{oxygen, N}_2} = V_{\text{tg, N}_2}$$

(9-3)

Carbon balance of COREX process:

$$\sum_i m_i \times w(C)_i + (V_{recg, CO} + V_{recg, CO_2} + V_{recg, CH_4}) \times \frac{12}{22.4}$$

$$= m_{HM} \times w[C]_{HM} + m_{dust} \times w(C)_{dust} + (V_{tg, CO} + V_{tg, CO_2} + V_{tg, CH_4}) \times \frac{12}{22.4}$$

$$(9-4)$$

In the equations, i denotes the raw materials and fuels; m_i represents the mass of material for producing 1t hot metal (kg/t); $w(j)_i$ denotes the mass fraction of j in i material; m_{HM} is the mass of hot metal (kg/t); m_{slag} denotes the mass of slag (kg/t); m_{dust} denotes the mass of dust (kg/t); $w[Fe]_{HM}$, $w[C]_{HM}$ denote the mass fraction of Fe, C in hot metal; $w(FeO)_{slag}$ denotes the mass fraction of FeO in slag; $w(Fe)_{dust}$, $w(C)_{dust}$ denote the mass fraction of Fe, C in dust; V_{oxygen} denotes the volume of oxygen for producing 1t hot metal (m³/t); x_{oxygen, N_2} denotes the volume fraction of nitrogen in oxygen (m³/t); V_{tg, N_2}, $V_{tg, CO}$, V_{tg, CO_2}, V_{tg, CH_4} denote the volume of N_2, CO, CO_2, CH_4 in top gas of reduction shaft (m³/t); $V_{recg, CO}$, V_{recg, CO_2}, V_{recg, CH_4} denotes the volume of CO, CO_2, CH_4 in recycle gas (m³/t).

Carbon balance of melter-gasifier:

$$m_{C(coke)} + m_{C(coke\ powder)} + m_{C(coal)} + m_{C(bdrs)} + m_{C(recycling\ dust)} + V_{recg, CH_4} \times \frac{12}{22.4}$$

$$= m_{C(combustion)} + m_{C(DR)} + m_{C(XO)} + m_{C([C])} + m_{C(recg)} + m_{C(dust\ in\ gg)} \quad (9-5)$$

where, $m_{C(coke)}$ is the amount of carbon in coke (kg/t); $m_{C(coke\ powder)}$ is the amount of carbon in coke powder (kg/t); $m_{C(coal)}$ is the amount of carbon in coal (kg/t); $m_{C(bdrs)}$ is the amount of carbon in the burden discharged from reduction shaft (kg/t); $m_{C(recycling\ dust)}$ is the amount of carbon in recycling dust (kg/t); $m_{C(recycling\ dust)}$ is the amount of carbon in recycling dust (kg/t); V_{recg, CH_4} is the volume of methane in recycle gas (m³/t); $m_{C(combustion)}$ is the amount of carbon consumed by combustion (kg/t); $m_{C(DR)}$ is the amount of carbon consumed by direct reduction (kg/t); $m_{C(XO)}$ is the amount of carbon consumed by non-ferrous oxide direct reduction (kg/t); $m_{C([C])}$ is the amount of carbon consumed by hot metal carburization (kg/t); $m_{C(recg)}$ is the amount of carbon consumed by recycle gas at tuyere (kg/t); $m_{C(dust\ in\ gg)}$ is the amount of carbon in dust of generator gas (kg/t).

Table 9-1 The chemical composition of raw materials (mass percent, %)

Raw material	SiO_2	Al_2O_3	CaO	MgO	TFe	FeO	S	P_2O_5	TiO_2	MnO	Ignition loss
Sinter	6.75	0.66	12.90	2.07	53.06	8.57	0.51	0.10	0.20	0.37	—
Pellet A	6.90	1.92	1.95	1.57	62.80	3.76	0.02	0.03	0.20	0.28	—

Table 9-1　The chemical composition of raw materials　(mass percent,%)

Raw material	SiO$_2$	Al$_2$O$_3$	CaO	MgO	TFe	FeO	S	P$_2$O$_5$	TiO$_2$	MnO	Ignition loss
Pellet B	6.25	1.88	2.41	1.46	60.40	2.87	0.03	0.07	0.27	0.25	—
Flux	0.47	—	54.20	0.43	—	—	0.01	—	—	—	44.86

Table 9-2　The chemical composition of fuels　(mass percent,%)

Fuel	Fixed Carbon	SiO$_2$	Al$_2$O$_3$	CaO	MgO	TFe	P$_2$O$_5$	H$_2$O	S
Coke A	81.84	5.99	2.52	1.20	0.35	1.00	0.05	4.60	0.50
Coke B	81.55	4.31	2.61	1.28	0.72	0.86	0	5.60	0.55
Coke nut	84.94	5.45	3.48	0.77	0.23	1.56	0	0.80	0.81
Coke powder	67.64	10.99	5.42	1.46	0.35	0.88	0.28	6.06	0.59
Lump coal	48.00	3.58	1.57	0.49	0.21	0.25	0	12.50	0.38

Table 9-3　The chemical composition of hot metal　(mass percent,%)

Type	Fe	C	Si	S	P	Mn	Ti
Content	93.27	4.66	1.70	0.02	0.09	0.19	0.07

9.1.3　The Top Gas Recycling Process

Generally, in the process of top gas recycling, the CO_2 in the gas needs to be removed to improve the reducibility and calorific value of the gas. For the top gas of reduction shaft, the physical absorption method (Selexol) is the most commonly selected solution among various CO_2 removal technologies. This method can remove 90% of CO_2 in the top gas of reduction shaft. Moreover, the energy consumption for CO_2 removal should be taken into account. The energy consumption of physical absorption method (Selexol) is 0.97 GJ/t CO_2 captured. For the whole process, this energy consumption should be included in the fuel rate. In addition, it is necessary to study the condition that the top gas without CO_2 removal is injected into the tuyere of melter-gasifier for comparison.

9.1.4　Nitrogen Accumulation

In the COREX process with top gas recycling, the nitrogen does not take part in any chemical reactions in furnace, some nitrogen is recycled back into the furnace with the recycle gas, and the rest of nitrogen is discharged as by-product gas. Thus, the unsteady recycling process would lead to the accumulation of N_2 in the top gas and recycle gas. However, there is a dynamic equilibrium that can stabilize the nitrogen content. In other words, when the nitrogen in the system is accumulated to a certain amount, the input of nitrogen is the same as the output of nitrogen, and the nitrogen in the system will not continue to be accumulated. The input and output chart of the nitrogen for the COREX system is shown in Fig. 9-4.

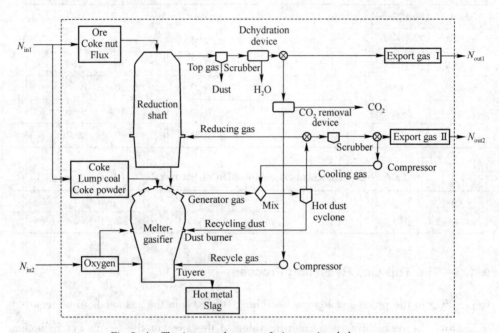

Fig. 9-4 The input and output of nitrogen in whole system

Input of N_2:

$$N_{in1} = \frac{22.4}{28} \times [m_{coal} \times w(N)_{coal} + m_{coke} \times w(N)_{coke} +$$
$$m_{coke\ nut} \times w(N)_{coke\ nut} + m_{coke\ powder} \times w(N)_{coke\ powder}] \quad (9-6)$$
$$N_{in2} = V_{oxygen} \times x_{oxygen,\ N_2} \quad (9-7)$$
$$N_{input} = N_{in1} + N_{in2} \quad (9-8)$$

Output of N_2:

$$N_{out1} = V_{ex1} \times x_{ex1N_2} \quad (9-9)$$
$$N_{out2} = V_{ex2} \times x_{ex2N_2} \quad (9-10)$$
$$N_{output} = N_{out1} + N_{out2} \quad (9-11)$$

Accumulation of N_2:

$$N_{accum} = N_{input} - N_{output} \quad (9-12)$$

In the above equations, m_{coal}, m_{coke}, $m_{coke\ nut}$, $m_{coke\ powder}$ is the mass of coal, coke, coke nut, coke powder (kg/t); $w(N)_{coal}$, $w(N)_{coke}$, $w(N)_{coke\ nut}$, $w(N)_{coke\ powder}$ denote the mass fraction of N in coal, coke, coke nut, coke powder; V_{oxygen} denotes the volume of oxygen (m³/t); x_{oxygen,N_2} denotes the volume fraction of nitrogen in oxygen; V_{ex1}, V_{ex2} denote the volume of export gas I, export gas II (m³/t); x_{ex1,N_2}, x_{ex2,N_2} denote the volume fraction of nitrogen in export gas I and export gas II respectively; N_{accum} denotes the accumulation of nitrogen (m³/t).

9.1.5 Calculation Method of CO₂ Emissions

The calculation method of the World Steel Association (WSA) for CO_2 emissions can be used in this paper. In this calculation method, CO_2 emissions equal the direct emission plus the indirect emission and minus the deductible carbon emission. The direct emission is the CO_2 emitted by the consumption of fuels and fluxes in the production process. The indirect emission is the CO_2 emitted by raw materials, fuels and energy in their own production processes. The deductible carbon emission is the amount of CO_2 emissions deducted from the sale or reuse of by products. The emission factors of WSA are the international average, and some emission factors are not applicable to China. Therefore, the measured emission factors are used in this paper, and the CO_2 emissions of COREX process and the CO_2 emissions of ironmaking system are calculated respectively. The CO_2 emissions of ironmaking system include the indirect emission of the upstream processes such as pelletizing and coking

9.2 Results and Discussion

Fig. 9-5 shows the mass flow of COREX process under the actual operation conditions. Under steady state, the consumptions of ore, fuel and coke (including coke, coke nut and coke powder) for the production of 1t hot metal are 1593.95kg/t, 830.74kg/t and 579.94kg/t, respectively. The effect of gas injection on the main operation parameters of COREX process is compared under the same smelting rate. The three cases are as follows: case 0 is the actual operation conditions, in other words, no recycle gas is injected; case 1 is that the top gas without CO_2 removal is injected at the tuyere, and the volume of recycle gas accounts for 5% of the total blast volume (oxygen+ gas); case 2 is that the gas after CO_2 removal is injected at the tuyere, and the volume of the recycle gas accounts for 5% of the total blast volume (oxygen+gas). The mass flows of case 1 and case 2 are shown in Fig. 9-6 and Fig. 9-7, respectively.

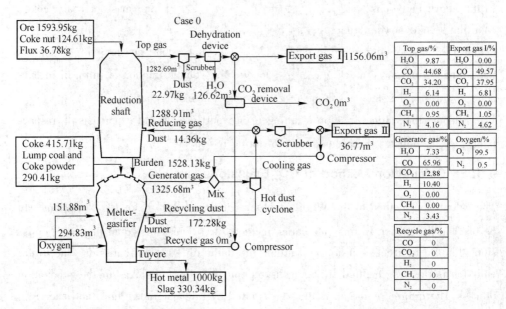

Fig. 9-5 The mass flow of COREX process under the actual operation conditions (case 0)

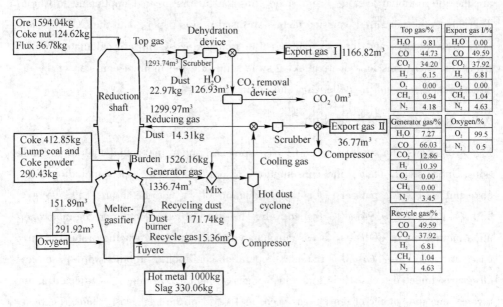

Fig. 9-6 The mass flow of COREX process for injecting the recycle gas without CO_2 removal (case 1)

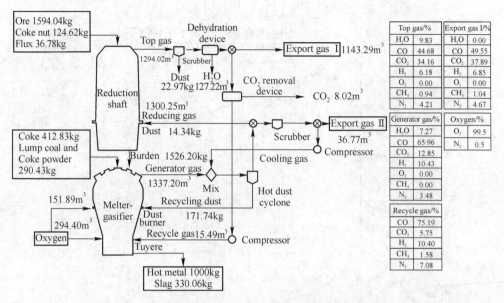

Fig. 9-7 The mass flow of COREX process for injecting the recycle gas after CO_2 removal (case 2)

9.2.1 Effect on Theoretical Combustion Temperature

Under the assumption that the combustion in the tuyere raceway is an adiabatic process, all the heat released by burning carbon to produce CO is used to heat the gas in the tuyere raceway, and the theoretical combustion temperature is the temperature that the gas can reach theoretically. The appropriate theoretical combustion temperature is good for the smooth operation, so the theoretical combustion temperature is an important parameter of COREX furnace. As can be seen from Fig. 9-8, no matter which kind of recycle gas is injected, the theoretical combustion temperature is decreased with the increase of recycle gas volume. Part of the reason is that the CO_2 in the gas will react with the carbon in the fuel to consume some heat. The other part is that the reduction of oxygen injected into the tuyere raceway reduces the amount of heat generated by carbon combustion. It can also be seen from the figure that the injection of RG-1 has the largest decrease in the theoretical combustion temperature. The main reason is that the content of CO_2 in the RG-1 is relatively high, so the amount of solution loss in the tuyere raceway is relatively large, thus the heat consumption is relatively large. Compared with injecting the RG-2, the decrease of theoretical combustion temperature by injecting the RG-3 is smaller. The reason is that the physical heat brought by the RG-3 is relatively high.

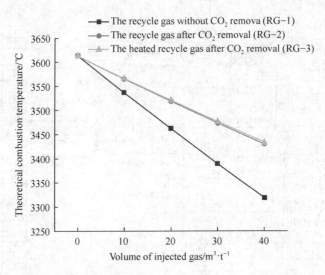

Fig. 9-8 Effect on theoretical combustion temperature

9.2.2 Effect on Dome Temperature

The dome temperature of melter-gasifier should be kept within a certain range to keep the smooth operation of COREX furnace. If the dome temperature of melter-gasifier is too low, the volatile matter in coal can not be completely cracked, and the gas pipeline may be blocked due to the formation of tar. In addition, if the dome temperature is too high, not only a lot of heat will be taken away by the gas, but also the refractory material and equipment in the free board may be damaged by high temperature gas. The production practice indicates that the dome temperature of melter-gasifier must be controlled at about 1050℃, and the appropriate dome temperature range is between 1000℃ and 1100℃. As can be seen from Fig. 9-9, the different types of recycle gas have different effects on the dome temperature. When the RG-1 is injected, the dome temperature gradually decreases with the increase of recycle gas volume. If the volume of recycle gas reaches or exceeds 30m³/t, the dome temperature will be lower than 1000℃, which will affect the smooth operation. With the increase of recycle gas volume, there is no obvious change in dome temperature when the RG-2 is injected. Therefore, injection of the RG-2 will not affect the stable operation of COREX furnace. When the RG-3 is injected, the dome temperature rises slightly with the increase of recycle gas volume. And the injection of the RG-3 has little effect on the dome temperature. However, if the gas temperature is greatly increased or the gas volume is greatly increased, the dome temperature may be too high, which may affect the stable operation of melter-gasifier.

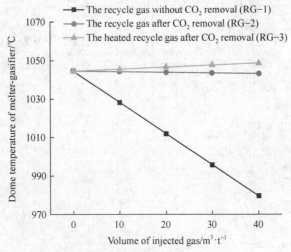

Fig. 9-9   Effect on dome temperature

9.2.3 Effect on Fuel Rate

It can be seen from Fig. 9-10 that if the energy consumption of CO_2 removal and heating of recycle gas is not took into account, the fuel rate can be decreased by injecting the three kinds of recycle gas with the increase of recycle gas volume. The effect of injecting the RG-1 and the RG-2 on fuel rate is very close, while the decrease of fuel rate by injecting the RG-3 is relatively large.

The energy consumption of 0.97 GJ is needed to remove 1 ton of CO_2 by physical adsorption (Selexol). If the energy consumption for CO_2 removal by physical absorption method (Selexol) is converted into coal rate, about 0.11kg coal is consumed to purify 1m^3 top gas of reduction shaft. When burning gas to heat the recycle gas, in order to increase the temperature of the 1m^3 recycle gas by 100℃, it is needed to burn about 0.02m^3 of the gas without CO_2 removal. It is considered that the integrated fuel rate is equal to the fuel rate plus the consumption of coal for CO_2 removal. It can be seen from Fig. 9-11 that the three kinds of recycle gas can reduce the integrated fuel rate with the increase of recycle gas volume. The results show that the decrease of integrated fuel rate is relatively small when injecting the RG-2, and the decrease of integrated fuel rate is relatively large when injecting the RG-3. The injection of RG-1 has the largest reduction on the integrated fuel rate. Although purifying and heating the recycle gas can reduce the fuel rate of COREX process, it also consumes energy to remove CO_2 and heat gas. Therefore, the overall decrease of integrated fuel rate by using these two methods is relatively small. Therefore, the integrated fuel rate can be significantly reduced by injecting a large amount of the RG-2 or the RG-3 in the tuyere.

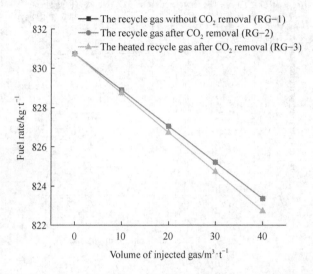

Fig. 9-10 Effect on fuel rate

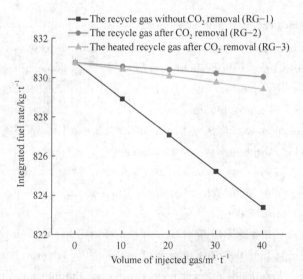

Fig. 9-11 Effect on integrated fuel rate

9.2.4 Effect on CO_2 Emissions

It can be seen from Fig. 9-12 and Fig. 9-13 that the injection of the three kinds of recycle gas can not only reduce the CO_2 emission of COREX process, but also can reduce the CO_2 emission of ironmaking system. For the CO_2 emission of COREX process, the injection of the RG-2 and the RG-3 can significantly reduce the CO_2 emission of COREX process. However, the emission reduction effect of the RG-2 is relatively small. And the results show that injection of the RG-1 has the smallest emission reduction effect. The CO_2 emission of COREX process does not include the CO_2 emission of other processes. Therefore, in order to comprehensively evaluate the CO_2 emission of COREX process combined with top gas recycling, the CO_2 emission of ironmaking system

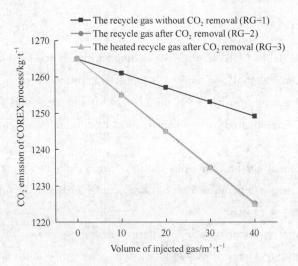

Fig. 9-12　Effect on CO_2 emission of COREX process

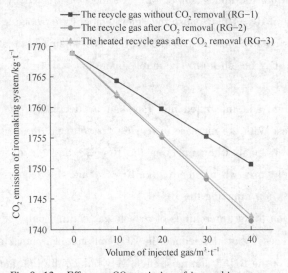

Fig. 9-13　Effect on CO_2 emission of ironmaking system

should be compared. For the CO_2 emission of ironmaking system, the injection of the RG-2 has the largest emission reduction effect, the emission reduction effect of the RG-3 is slightly lower, and the emission reduction effect of the RG-1 is the smallest. The reason is that the recycle gas is heated by burning the gas without CO_2 removal, and this part of gas will also produce CO_2 emission after combustion. Therefore, injection of the RG-2 can minimize the CO_2 emission of ironmaking system.

9.3 Summary

A mathematical model of COREX process combined with top gas recycling technology is developed based on material and energy balance. The influence of injecting the three kinds of recycled gas at the tuyere on the operation parameters and CO_2 emissions of COREX process is investigated by the mathematical model. Following conclusions are obtained from the above results.

(1) The theoretical combustion temperature is greatly affected by gas injection. The theoretical combustion temperature can be significantly reduced by injecting the three kinds of recycle gas at tuyere.

(2) The different types of recycle gas have different effects on the dome temperature. With the increase of recycle gas volume, the dome temperature is significantly reduced by injecting the RG-1, and the dome temperature is slightly increased by injecting the RG-3, while the dome temperature is almost unaffected by injecting the RG-2. A small amount of RG-1 may cause the dome temperature to be too low. If the recycle gas temperature is greatly increased or the recycle gas volume is greatly increased, the dome temperature may be too high. The stable operation is not affected by injecting the RG-2.

(3) The fuel rate and integrated fuel rate can be decreased by injecting the three kinds of recycle gas. With the increase of recycle gas volume, the decrease of integrated fuel rate is relatively small when injecting the RG-2, and the decrease of integrated fuel rate is relatively large when injecting the RG-3, and the decrease of integrated fuel rate is the largest when injecting the RG-1.

(4) The injection of the three kinds of recycle gas can not only reduce the CO_2 emission of COREX process, but also reduce the CO_2 emission of ironmaking system. Moreover, the injection of the RG-3 can most effectively reduce the CO_2 emission of COREX process. The injection of the RG-2 can minimize the CO_2 emission of ironmaking system.

References

[1] Wang H T, Chu M S, Guo T L, et al. Mathematical simulation on blast furnace operation of coke oven gas injection in combination with top gas recycling [J]. Steel Research International, 2016, 87: 539~549.

[2] Zhang W, Xue Z L, Zhang J H, et al. Medium oxygen enriched blast furnace with top gas recycling strategy [J]. Journal of Iron and Steel Research International, 2017, 24: 778~786.

[3] Qu Y X, Zou Z S, Xiao Y P. A comprehensive static model for COREX process [J]. ISIJ international, 2012, 52: 2186~2193.

[4] Lee S C, Shin M K, Joo S, et al. A development of computer model for simulating the transport phenomena in COREX melter-gasifier [J]. ISIJ international, 1999, 39: 319~328.

10 Influence of Cohesive Zone Shape on Solid Flow in COREX Melter Gasifier by Discrete Element Method

Melter gasifier is similar to the blast furnace (BF) that is also a gas-solid counter-current reactor. Direct reduction iron (DRI) and coal (or coke) particles are mixing charged from the upper of melter gasifier, and the pure oxygen with room temperature is blown from tuyeres. DRI particles are reduced during descending and many physical changes and chemical reactions between each phase are undergoing in there. Its stable operation is closely related with the smoothly descending of burden. The terrible phenomena like tunnel and hanging will happen once the normal descending condition is broken. Thus, it is necessary to study the burden behavior in the COREX melter gasifier.

In this presentation, based on the principle of DEM, a slot model of COREX melter gasifier has been established. Using the proposed model, the influence of cohesive zone shape and the relationship among the mass distribution, particle velocity and normal force distribution are discussed to clarify the total burden descending in melter gasifier by DEM at a microscopic level.

10.1 Simulation Condition

In the present calculation, the solid flow behavior in COREX 3000 melter gasifier was simulated by DEM. The geometry of melter gasifier used and the three kinds of cohesive zone proposed in this simulation are shown in Fig. 10-1, which is the same as the experiment device used by Zhou. Two tuyeres are set at 100mm from the hearth bottom level at each side. The shape of raceway is assumed to be spherical shape with 20mm in radius. The particles that flow into raceway disappear at the specified interval. The three kinds of cohesive zones, which are unknown whether they are existent or not in melter gasifier, are assumed to be inverse 'V' in the present simulation. Cohesive zone (a) called as base cohesive zone is a symmetric cohesive zone with 350mm in height over hearth bottom level. Cohesive zone (b) is a high cohesive zone with 375mm in height over hearth bottom level. Cohesive zone (c) is a biased cohesive zone with 80mm biased distance from center of melter gasifier. The height of biased cohesive zone is

350mm over hearth bottom level. The reduction of DRI particles in this region is considered by their shrinking to small particles. The transient diameter of DRI particles in the cohesive zone is determined by equation (10-1).

$$d_{DRI} = d_{0,DRI} - (d_{0,DRI} - d_{c,DRI})(Z_{top} - Z_p)/(Z_{top} - Z_{bot}) \quad (10-1)$$

where $d_{c,DRI}$ is the critical diameter of DRI particles when leaving the cohesive zone (set to $0.4d_{0,DRI}$ and $d_{0,DRI}$ is the original diameter of DRI particles), and Z_{top}, Z_{bot} and Z_p are vertical distances to the hearth bottom wall as shown in Fig. 10-1(a). When DRI particles leave the cohesive zone, they become liquid, which is however not considered in this work focused on solid flow behavior. The downward flow of burden due to gravity is driven mainly by the combustion of coke particle in raceway and disappearance of DRI in cohesive zone.

The procedure for the numerical experiment is described as follows. 16000 spherical and uniform particles are firstly generated in the model melter gasifier, as shown in Fig. 10-2, without considering gas blast, hearth liquid, and particle discharging from the raceway. After the structure of the packed bed is established, particles are then removed from the raceway at a rate to represent the coke combustion. As is well known, the melter gasifier employs the mixing charging pattern, which is different from the traditional BF. To simplify this charging process, the model is divided into three zones to charge particles with the mixing charging pattern along the radial direction, where the DRI to coke and coal volume ratio are 2 : 1, 1 : 1 and 2 : 1, respectively, corresponding to ①, ② and ③, that is shown in Fig. 10-3, in which the black particles denote DRI and the gray ones denote coke or coal. The material properties in the simulation are listed in Table 10-1.

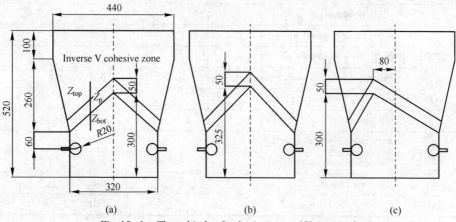

Fig. 10-1 Three kinds of cohesive zone (Unit: mm)
(a) Base cohesive zone; (b) High cohesive zone; (c) Biased cohesive zone

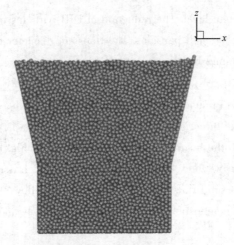

Fig. 10-2 Initial packing state in the melter gasifier model

Fig. 10-3 The distribution of DRI and coke or coal

Table 10-1 Simulation parameters

Variables	Value
Particle shape	Sphere
Particle number N	16000
Particle density ρ_p/kg·m^{-3}	1100/4000
Particle diameter d_p/mm	10
Particle-particle sliding frictional coefficient $\mu_{s,pp}$	0.5
Particle-particle rolling frictional coefficient $\mu_{s,pp}$	0.0005
Particle-wall sliding frictional coefficient $\mu_{r,pw}$	0.5
Particle-wall rolling frictional coefficient $\mu_{r,pw}$	0.0005
Young's modulus of particle E_p/Pa	2160000
Young's modulus of wall E_w/Pa	2160000
Poisson's ratio of particle v_p	0.3
Poisson's ratio of wall v_w	0.3
Time step Δt/s	1.0×10^{-4}

Note: It is assumed that the walls have the same properties as the particles in this presentation.

10.2　Results and Discussion

10.2.1　Influence of Cohesive Zone Shape on the Mass Distribution

Fig. 10-4 shows the mass distribution under different cohesive zones in this present model, and different colors denote different mass. It can be seen from the figures that the kinds of cohesive zones have little influence on the burden structure in the upper part. However, the shape of cohesive zone slightly causes disturbance of burden structure in lower part of shaft due to the narrow distance between the cohesive zone and wall. Especially under the condition of biased cohesive zone of case (c), the shape of

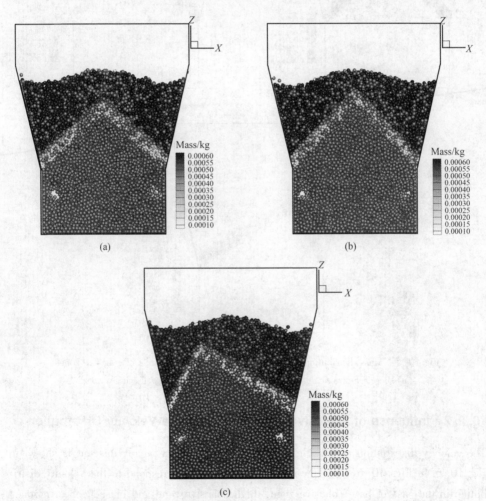

Fig. 10-4　Mass distribution under three kinds of cohesive zones
(a) Base cohesive zone; (b) High cohesive zone; (c) Biased cohesive zone

burden structure on both sides becomes asymmetric. Fig. 10-5 shows the mass distribution of cohesive zone, and what can be seen from all the figures is that the DRI particles size decreases gradually from the upper part of cohesive zone to the lower part.

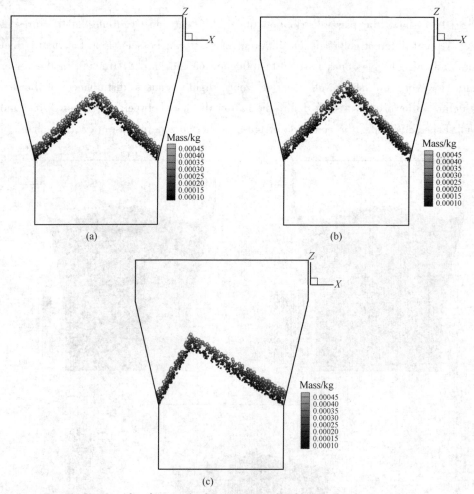

Fig. 10-5 Mass distribution of cohesive zone under three kinds of cohesive zones
(a) Base cohesive zone; (b) High cohesive zone; (c) Biased cohesive zone

10. 2. 2 Influence of Cohesive Zone Shape on the Velocity Distribution

The velocity distribution of particle in the 2D slot model at steady descent is shown in Fig. 10-6. In Figs. 10-6(a), 10-6(b) and 10-6(c) correspond to the velocity distribution in the case of base cohesive zone, high cohesive zone and biased cohesive zone, respectively. There are several interesting features regarding these figures. It is observed that the velocity field can be divided into several characteristic regions: (1) funnel flow region with high velocity shown in deep gray; (2) coke moving zone with medium velocity shown in light gray; (3) deadman with stagnant coke velocity shown in deep black; (4) quai-stagnant zone adjacent to the deadman shown in light black. It is observed that deadman exists in the lower central part of blast furnace and the shape of deadman is almost not influenced by the cohesive zone shape in every case. The shape of coke moving zone corresponds to the shape of cohesive zone. An asymmetric medium velocity region is formed in lower part under the condition of biased cohesive zone. High cohesive zone enlarges the medium velocity region to upper part. Accordingly, it is estimated that the shape of deadman is almost independent on the cohesive zone and coke moving zone is influenced by the melting behavior of burden.

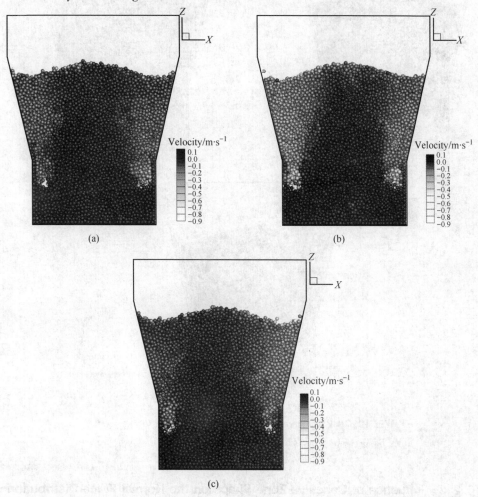

Fig. 10-6 Velocity distribution of particles at steady descent
(a) Base cohesive zone; (b) High cohesive zone; (c) Biased cohesive zone

Fig. 10-7 shows the particle velocity vectors corresponding to Fig. 10-6. It can be also seen from this figure that the high velocity region is located near the wall region and the raceway, and the low velocity region is located under the lower part of cohesive zone. Under the condition of biased cohesive zone of case (c), its velocity distribution becomes asymmetric. Fig. 10-8 shows the particle velocity vectors in cohesive zone that also presents the same phenomenon.

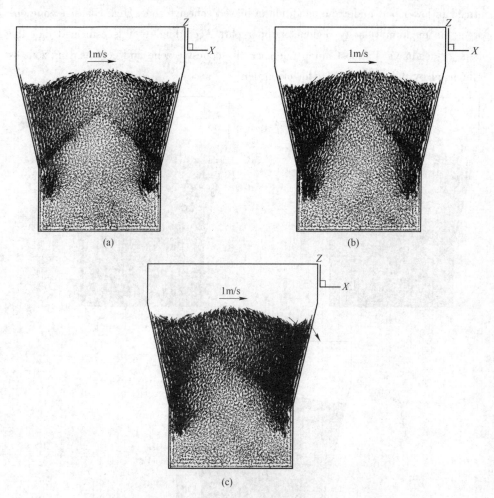

Fig. 10-7 Particle velocity vectors corresponding to Fig. 10-6
(a) Base cohesive zone; (b) High cohesive zone; (c) Biased cohesive zone

10.2.3 Influence of Cohesive Zone Shape on the Normal Force Distribution

Fig. 10-9 shows the distribution of the normal force in which the length of the arrows re-

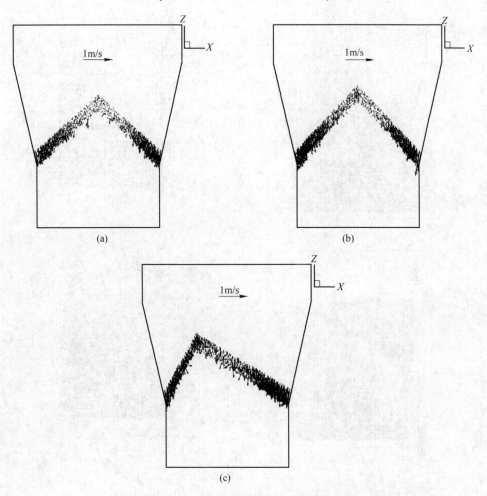

Fig. 10-8　Particle velocity vectors of cohesive zone
(a) Base cohesive zone; (b) High cohesive zone; (c) Biased cohesive zone

presents the magnitude of the force between two particles under three kinds of cohesive zones. It can be observed that the particles in the lower central bottom experience large normal forces. This is because these particles need to support the particles above them. On the contrary, particles exhibit weak force network around the raceway and in the fast flow zone. This is because particles in these regions flow fast and there exist more voids and disconnections among particles. It is also observed that the distribution of normal force in lower shaft is not influenced by the cohesive zone shape. Obviously, the phenomena are in general agreement with our common sense. However, microscopic information obtained from DEM can help develop better understanding of solid flow behavior inside melter gasifier.

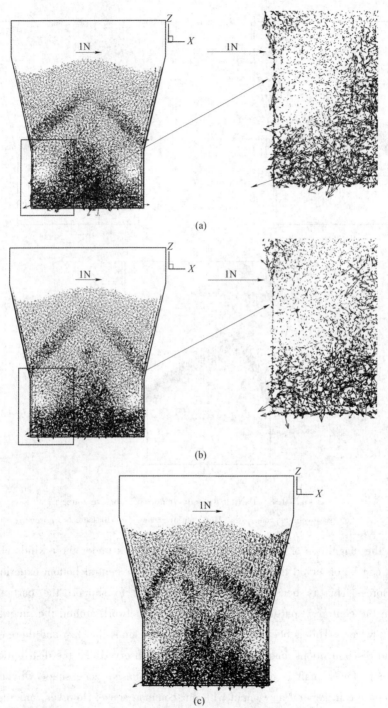

Fig. 10-9 The distribution of the normal force under three kinds of cohesive zone
(a) Base cohesive zone; (b) High cohesive zone; (c) Biased cohesive zone

10.2.4 Influence of Cohesive Zone Shape on the Normal Force Distribution

The flow structure of particles can be examined in terms of porosity which has been widely used in the study of particle packing. As used elsewhere, porosity is here treated as a local mean property which is obtained by dividing the particle bed into a series of body-fitted cells, and then calculating porosity for each cell. The spatial distribution of porosity is related to the flow of particles, for example, the region with high porosity must be from the unconfined motion of particles. It is also closely related to permeability distribution which governs the flow of gas, liquid and powder, and is critical to smooth and stable COREX melter gasifier operation. Under the 2D slot model conditions, a typical spatial porosity distribution is shown in Fig. 10-10. It reveals that the whole packed bed can be divided into four main regions: (1) lowest porosity region (0.235 ~ 0.312) located in the cohesive zone. That is because DRI particles there exist softening melting and adhesive; (2) low porosity region (0.312 ~ 0.426) located in the hearth and deadman, because particles there need to form a relatively dense packing to support the rest of the bed; (3) medium porosity region (0.426 ~ 0.503) which mainly corresponds to the plug flow; (4) high porosity region (0.503 ~ 0.579, or even higher in the raceway) which corresponds to the raceway and converging flow zones where particles have high velocities. With the increase of cohesive zone position, the low porosity region located in the root of cohesive zone increases. The porosity distribution becomes asymmetric in the case of biased cohesive zone.

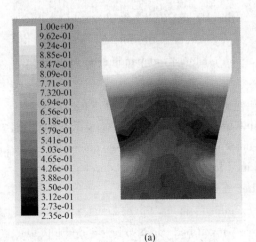
(a)

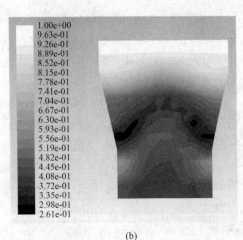
(b)

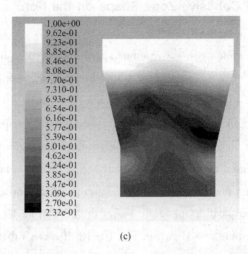

(c)

Fig. 10-10　Porosity distribution under different cohesive zones
(a) Base cohesive zone; (b) High cohesive zone; (c) Biased cohesive zone

10.3　Conclusions

Three-dimensional discrete element method (DEM) has been employed to investigate the influence of cohesive zone shape on solid flow in a slot melter gasifier. The following results were obtained.

(1) The mass distribution in upper melter gasifier is not affected by the cohesive zone shape. Asymmetric burden structure is formed in the case of biased cohesive zone and the DRI particles size in cohesive zone decreases gradually from the upper part of cohesive zone to the lower part.

(2) The velocity distribution of particle in the 2D slot model is also studied under three kinds of cohesive zones. There exist four characteristic regions just like the BF. And it is estimated that the shape of deadman is almost independent on the cohesive zone.

(3) The particles in the lower central bottom experience large normal force to support the particles above them. And particles exhibit weak force network around the raceway and in the fast flow zone. Although the phenomena are in general agreement with our common sense, microscopic information obtained from DEM can help develop better understanding of solid flow behavior inside melter gasifier.

(4) The porosity distribution is also examined under three kinds of cohesive zones. Like the velocity distribution, the whole packed bed can also be divided into four main regions. With the increase of cohesive zone position, the low porosity region

located in the root of cohesive zone increases. And the porosity distribution becomes asymmetric in the case of biased cohesive zone.

References

[1] Natsui S, Ueda S, Oikawa M, et al. Optimization of physical parameters of discrete element method for blast furnace and its application to the analysis on solid motion around raceway [J]. ISIJ International, 2009, 49: 1308~1315.

[2] Zhu H P, Zhou Z Y, Yang R Y, et al. Discrete particle simulation of particulate systems: theoretical developments [J]. Chemical Engineering Science, 2007, 62: 3378~3396.

[3] Li H F, Luo Z G, Zou Z S, et al. Mathematical simulation of burden distribution in COREX melter gasifier by discrete element method [J]. Journal of Iron and Steel Research International volume, 2012, 19: 36~42.

11 Influence of Burden Distribution on Temperature Distribution in COREX Melter Gasifier

In Melter gasifier, the gas distribution mainly affected by the burden distribution is a main factor that influences the furnace temperature. Proper distribution of burden materials can improve bed permeability, wind acceptance, and efficiency of gas utilization. At present, DRI, lump coal and coke mixed charging pattern is being employed in COREX melter gasifier, which is significantly different from BF. Therefore, it is necessary to study the influence of burden distribution on temperature distribution in COREX melter gasifier.

The overall purpose of this study was to investigate the influence of burden distribution including radial distribution of DRI to lump coal and coke volume ratio, coke charging location, as well as coke charging amount and coke size on the temperature distribution in COREX melter gasifier. Furthermore, the gas flow could also be analyzed, which can provide references for optimizing the burden distribution and improving the PCI rates.

11.1 Experimental

11.1.1 Experimental Apparatus

A two-dimensional 1/30 scale thermal dynamic model of COREX 3000 melter gasifier has been developed. Fig. 11-1 shows a schematic diagram of the experimental apparatus, including the main apparatus, discharging system at the hearth bottom, hot air supply system and temperature acquisition system.

The main apparatus are mainly made of stainless steel. Especially, the front side of the apparatus is made of heat-resistant glass to observe the descending and melting behaviors of quasi-coke and quasi-DRI. To enhance the heat preservation effect of the furnace, the display panel is covered by two-layer heat-resistant glass and where around the furnace is covered by thermal insulation material except for the display panel.

Discharging system at the hearth bottom mainly consists of two parts, mechanical transmission and furnace bottom baffle. The mechanical transmission can control the moving speed of the baffle, further adjust the discharging rate.

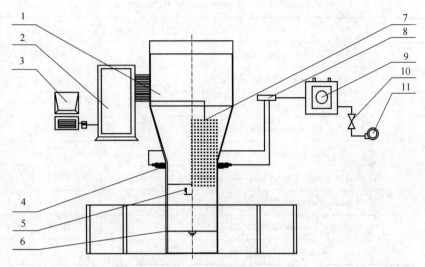

Fig. 11-1　Experimental apparatus

1—Two-dimensional melter gasifier model; 2—Temperature transmitter; 3—Computer; 4—Tuyere;
5—Mechanical transmission mechanism at the hearth bottom; 6—Furnace bottom baffle;
7—Thermocouples; 8—Gas distributor; 9—Air heater; 10—Flowmeter; 11—Air blower

The hot air supply system with the motor power 10 kW limits the blast temperature error to 0.1 ℃.

Temperature acquisition system consists of thermocouples, temperature transmitter and computer. The temperature can be measured by thermocouples which are installed at the back of the furnace. The temperature transmitter and computer are used to collect data and save them, respectively.

11.1.2　Experimental Conditions

In the experiment, the paraffin and the corn particles are used to simulate DRI, coke and lump coal, respectively. After the materials are mixed proportionally, the materials are charged into the model from the top to a certain height and then they will be discharged from the bottom. Moreover, the materials addition and excretion can be calculated according to the practical production and mass conservation law. Similarly, the blast volume is calculated by using the Froude number. According to the literature, the process of melting DRI, that the pure oxygen blasted into furnace reacts with lump coal and coke which can let out a large quantity heat to melt DRI, was simulated by using hot blast to melt the paraffin, not to consider the chemical reaction in the furnace. And further the blast temperature can be calculated according to mass conservation law and heat balance. The experimental parameters are listed in Table 11-1.

Table 11-1 Experimental Parameters

Parameter	Value
Bed height/mm	370
Charging materials temperature/℃	20
Discharging rate/r · min^{-1}	0.5
Blast temperature/℃	100
Blast volume/m^3 · min^{-1}	14
Radial distribution of relative DRI to lump coal and coke volume ratio	2 : 3, 1 : 1, 3 : 2
Amount of central coke charging/%	0, 8.7, 17.4
Amount of intermediate coke charging/%	0, 8.7, 17.4
Coke charging size/mm	3, 5

11.1.3 Experimental Procedures

The experimental procedures are as follows:

(1) The bed is packed, and thermocouples are properly located, according to the experimental parameters.

(2) Gas flow is set according to the experimental parameters and released until the blast temperature reaches steady state.

(3) The gas heater is set to the desired setpoint, and data logging of thermocouples is commenced, when gas flow is introduced to the apparatus.

(4) When the majority of paraffin has melted away, an apparent steady-state is achieved, or the chosen quench point has been reached, the gas heater is stopped followed by the gas flow.

(5) Data logging is stopped and the results are saved for later analysis.

(6) For the quench runs, the apparatus is allowed to cool to room temperature and then the bed is excavated to undertake next experiment.

11.2 Experimental Results and Discussion

11.2.1 Influence of Radial Distribution of Relative DRI to Lump Coal and Coke Volume Ratio on the Temperature Distribution

To optimize the burden distribution, the influence of radial distribution of relative DRI to lump coal and coke volume ratio on the temperature distribution in COREX melter gasifier was investigated. As is well known, coal is charged into the furnace by employing Gimbal distributor while DRI is charged into the furnace by employing DRI-flap distributor in actual production. Different radial distribution of DRI to lump coal and coke volume ratio can be obtained through adjusting these two distributors. In view of the complex problem of burden distribution in COREX practice and with the charging process being simplified in the laboratory, the radial direction of model is divided into two regions where each region employs the mixed burden distribution to charge the materials under total DRI to lump coal and coke volume ratio being constant. The radial distribution of DRI to lump coal and coke volume ratio is obtained according to the actual furnace operation conditions under some month, as shown in Fig. 11-2. The relative DRI to lump coal and coke volume ratio was calculated as the ratio of the DRI to lump coal volume ratio between 0 ~ 8cm and 8 ~ 16cm in the radial direction.

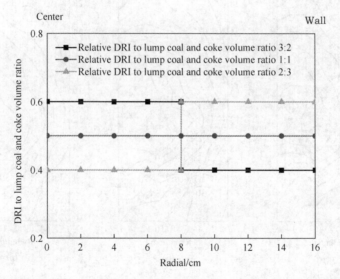

Fig. 11-2 The radial distribution of DRI to lump coal and coke volume ratio

Fig. 11-3 shows the influences of radial distribution of relative DRI to lump coal and coke volume ratio on the temperature distribution under mixed charging pattern in melter gasifier. Obviously, it can be clearly seen from the isothermal diagram that with the decrease of the radial distribution of relative DRI to lump coal and coke volume ratio, the temperature near the wall region decreases. The reason is that the DRI to lump coal and coke volume ratio is high in the wall region where it contains more DRI. Thus, it needs more heat to melt the DRI, and moreover, the increase of the content of DRI would result in low gas permeability in the wall region. What could also be found in Fig. 11-3 is that the temperature away from the wall region rises to some extent under the condition of the radial distribution of relative DRI to lump coal and coke volume ratio being equal to 1 : 1 and the isotherm near the wall region has greater temperature gradient. It is indicated that the gas flow near the wall region is also restrained. It can be inferred that this charging pattern is reasonable. Therefore, the following experiments of central coke charging and intermediate coke charging are performed under fixed radial distribution of relative DRI to lump coal and coke volume ratio being equal to 1 : 1.

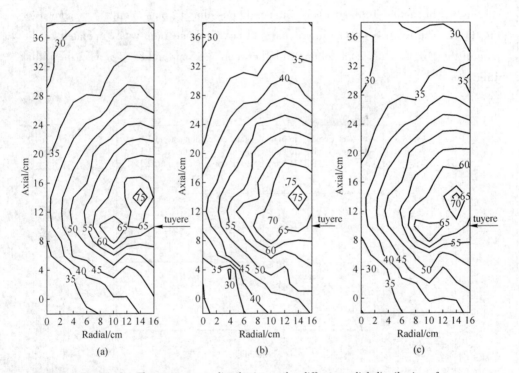

Fig. 11-3 The temperature distribution under different radial distribution of relative DRI/(lump coal and coke) volume ratio
(a) 3 : 2; (b) 1 : 1; (c) 2 : 3

11.2.2 Influence of Coke Charging Location on the Temperature Distribution

11.2.2.1 Influence of Central Coke Charging on the Temperature Distribution

For blast furnace, the coke charging amount is closely related with the condition of raw material, fuel and operation system. In this work, the coke charging amount is assumed to be 8.7% and 17.4% of the total coke charging. Fig. 11-4 shows the schematic diagram of coke charging location that the gray area represents the central coke charging and the black area represents the intermediate coke charging, respectively. Fig. 11-5 shows the influences of central coke charging on the temperature distribution in the furnace. Both the temperature with central coke charging is higher than that without central coke column. This may be attributed to the central coke charging in the furnace which allows the furnace to have better gas permeability. Furthermore, the temperature significantly increases with the increase of central coke charging amount. Also, as shown in Fig. 11-5, due to the isotherm having the high inclination, it is indicated that the peripheral gas flow is restrained by charging the central coke column.

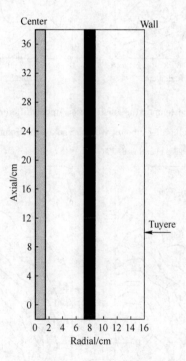

Fig. 11-4 Schematic of coke charging location
(The gray area represents the central coke charging, the black area represents the intermediate coke charging)

11.2.2.2 Influence of Intermediate Coke Charging on the Temperature Distribution

Fig. 11-6 shows the temperature distribution under different intermediate coke charging amount. With the increase of intermediate coke charging amount, the temperature near the wall region decreases while the temperature in the intermediate region increases. It is

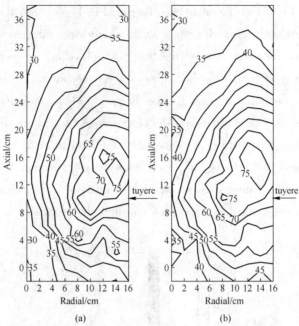

Fig. 11-5 The temperature distribution under different central coke charging amount with the coke size 3mm

(a) Central coke charging 8.7%; (b) Central coke charging 17.4%

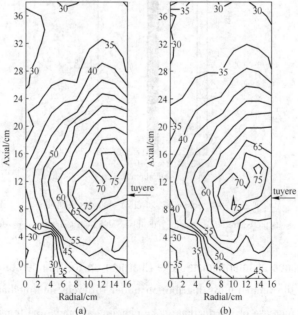

Fig. 11-6 The temperature distribution under different intermediate coke charging amount with the coke size 3mm

(a) Intermediate coke charging 8.7%; (b) Intermediate coke charging 17.4%

supposed that the gas flow presents two streams ascending in the furnace. Charging coke at intermediate can improve the gas flow at intermediate and reduce the wall erosion by gas flow.

11.2.3 Influence of Coke Size on the Temperature Distribution

In Fig. 11-7(a) and Fig. 11-7(b), it is obviously seen that the furnace temperature increases with the increase of coke size, respectively, compared with Fig. 11-5(a) and Fig. 11-5(b). This is associated with better gas permeability of packed bed, reducing the gas flow resistance. Similarly, the phenomenon that the temperature increases is also found by comparing Fig. 11-8(a) and Fig. 11-8(b) with Fig. 11-6(a) and Fig. 11-6(b), respectively. It could be inferred that the burden distribution with central and intermediate coke charging is effective to improve the PCI rates. Nevertheless, the gas utilization could be also lowered, that is necessary to be studied further.

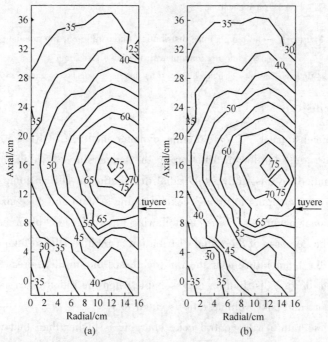

Fig. 11-7 The temperature distribution under different central coke charging amount with the coke size 5mm

(a) Central coke charging 8.7%; (b) Central coke charging 17.4%

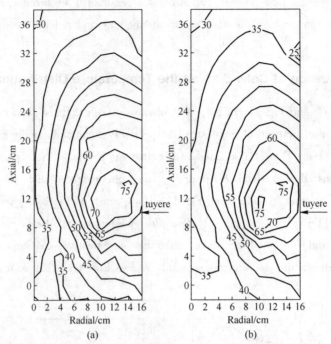

Fig. 11-8 The temperature distribution under different intermediate
coke charging amount with the coke size 5mm
(a) Intermediate coke charging 8.7%; (b) Intermediate coke charging 17.4%

11.3 Conclusions

Using a two-dimensional 1/30 scale thermal dynamic model of COREX 3000 melter gasifier, the influence of burden distribution on the temperature distribution in the melter gasifier was studied. In this study, the radial distribution of DRI to lump coal and coke volume ratio, coke charging location, coke charging amount and coke size were considered as operational parameters, and the following knowledge was obtained.

(1) With the decrease of the radial distribution of relative DRI to lump coal and coke volume ratio, the temperature near the wall decreases due to the increase of DRI in the wall region. The burden distribution is reasonable when the radial distribution of relative DRI to lump coal and coke volume ratio is equal to 1 : 1.

(2) The temperature with central coke charging is higher than that without central coke column. Furthermore, the temperature significantly increases with the increase of central coke charging amount.

(3) With the increase of intermediate coke charging amount, the temperature near the wall region decreases while the temperature in the intermediate region increases. It is

supposed that the gas flow presents two streams ascending in the furnace.

(4) The furnace temperature increases with the increase of coke size both under central coke charging and intermediate coke charging. It is inferred that controlling of the particle diameters of lump coal and coke can optimize the gas flow in the melter gasifier.

References

[1] Wang F, Bai C G, Yu Y W, et al. Influence of bed condition on gas flow in corex melter gasifier [J]. Ironmaking and Steelmaking, 2009, 36 (8): 590~596.

[2] Takahashi H, Komatsu N. Cold model study on burden behaviour in the lower part of blast furnace [J]. ISIJ International, 1993, 33 (6): 655~663.

[3] Takahashi H, Tanno M, Katayama J. Burden descending behaviour with renewal of deadman in a two dimensional cold model of blast furnace [J]. ISIJ International, 1996, 36 (11): 1354~1359.

Part IV

Simulation of Fine Particles Behavior in COREX

12 Experimental Study and Numerical Simulation of Dust Accumulation in Bustle Pipe of COREX Shaft Furnace with Areal Gas Distribution Beams

In COREX process, the generated gas from melter gasifier, enters the hot gas cyclone for dedusting. The large dust is removed and the reduction gas is considered roughly cleaned. A dust load of $20g/m^3$ is blasted into the shaft furnace through the bustle pipe. The initial gas distribution is complicated and tiny dust particles could be deposited in the packed bed, which would form a blockage near the slot. The blocked region may continue to grow, spreading into the bustle pipe. This directly affects the smoothness and performance of the shaft furnace (SF). The dust accumulation in bustle pipe of the SF has not been studied. There have been experimental and numerical studies that explain gas-fine flow and fine particle blockage in a packed bed, but the understanding of the gas-fine flow in the SF bustle pipe and the initial position of the dust accumulation remains unknown.

In this work, physical and mathematical modeling were performed for the gas-dust flow in bustle pipe of SF in order to better understand dust accumulation behavior. The initial chocking slots in bustle pipe were identified. The effect of the blast volume on dust accumulation was studied. These findings are expected to be useful for the control and optimization of the shaft furnace operation.

12.1 Experimental

The similarity theory states that the geometric similarity is initially guaranteed. A 1/20 scale semi-cylindrical model of the COREX-3000 shaft furnace was used to observe the dust accumulation phenomenon in the bustle pipe. Fig. 12-1(a) depicts the schematic diagram of the experimental apparatus. The shaft body was composed of perspex. Particles were charged into the model from the top to maintain a certain bed height. They were extracted from five screw dischargers. The rate of the particle extraction was controlled by induction motors. Powders were supplied from a dispenser at a constant rate. The powders were introduced to the gas stream before the inlet. Gas was added with a blower. The gas flow rate was measured with a flow meter. The bustle pipe of the cold model had two AGD inlets and forty slots, as shown in Fig. 12-1(b). The two-phase flow of the gas and the powders were initially blown into the bustle pipe. They then flowed into the SF through the AGD inlets and slots. A schematic diagram that illustrates the gas flow from the bustle into the furnace is shown in Fig. 12-1(c). The evolution of the dust accumulation was recorded with cameras in the quarter region of the bustle pipe.

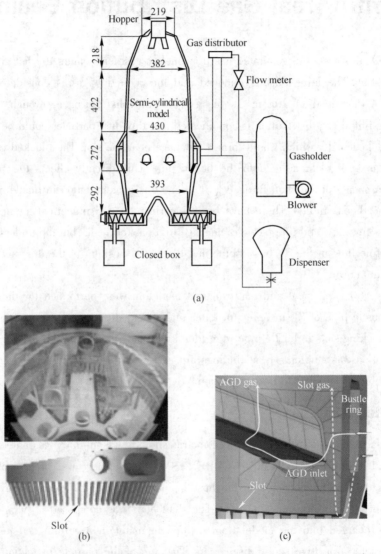

Fig. 12-1 A 1/20 scale semi-cylindrical model of the COREX-3000 shaft furnace
(a) Schematic diagram of the semi-cylindrical apparatus (Unit: mm);
(b) Outside view of the bustle pipe; (c) Schematic diagram of the gas distribution in the bustle pipe

The dynamic similarity between the model and the prototype was successful. This was performed by considering the various dominating forces, such as inertial, gravitational, and viscous forces. The Reynold number was correlated between the inertial force and the viscous force. These numbers should be set in the same state (laminar or turbulent). The inertial and the gravitational forces were primarily responsible for the fluid flow. These were taken into consideration when determining the model parameters. The Froude number was the most important non-dimensional parameter and was made equal in the prototype and in the model. The physical properties, particularly the repose angle of corn (diameter: 3mm, bulk density: 720kg/m^3, repose angle: 36°) were approximately equal to the average of the coke (bulk density: 500kg/m^3, repose angle: 43.5°) and the pellet (bulk density: 2000kg/m^3, repose angle: 32°). Grains of corn were used to simulate the burden materials. The glass beads (diameter = 90μm, density = 2200kg/m^3) had nearly the same gravity of the dust in the actual SF (75μm diameter) that were used in the current experiments.

12.2 Mathematical Model

The Ansys-Fluent 19.5 commercial software was used in the study. The dust motion was calculated by using the discrete particle model (DPM). The basic mathematical model equations that described the phenomena under examination are as follows:

Continuity equation:

$$\frac{\partial \rho}{\partial t} + \nabla \cdot (\rho \boldsymbol{v}) = S_m$$

Momentum equation:

$$\frac{\partial}{\partial t}(\rho \boldsymbol{v}) + \nabla \cdot (\rho \boldsymbol{v}\boldsymbol{v}) = \nabla \cdot (\mu \mathrm{grad}\boldsymbol{v}) + S_v$$

Turbulent kinetic energy k equation:

$$\frac{\partial}{\partial t}(\rho k) + \nabla \cdot (\rho k \boldsymbol{v}) = \nabla \cdot \left[\left(\mu + \frac{\mu_t}{\sigma_k}\right) \mathrm{grad} k\right] + G_k + G_b + \rho \varepsilon - Y_M + S_k$$

Turbulent energy dissipation rate ε equation:

$$\frac{\partial}{\partial t}(\rho \varepsilon) + \nabla \cdot (\rho k \boldsymbol{v}) = \nabla \cdot \left[\left(\mu + \frac{\mu_t}{\sigma_\varepsilon}\right) \mathrm{grad} \varepsilon\right] + C_{1\varepsilon} \frac{\varepsilon}{k}(G_k + C_{3\varepsilon} G_b) - C_{2\varepsilon} \rho \frac{\varepsilon^2}{k} + S_\varepsilon$$

The constants used in the k-ε model were: $C_{1\varepsilon} = 1.44$, $C_{2\varepsilon} = 1.92$, $C_\mu = 0.09$, $\sigma_k = 1.0$, and $\sigma_\varepsilon = 1.3$.

The trajectories of the dust were predicted by integrating a force balance on the dust. This force balance equated the dust inertia with the forces acting on the dust, and is shown as follows:

$$\frac{dv_p}{dt} = F_D(v - v_p) + \frac{g(\rho_p - \rho)}{\rho_p} + F$$

where, v is the gas velocity; v_p is the velocity of dust; ρ is the density of gas; ρ_p is the density of dust; F is an additional acceleration term. $F_D(v-v_p)$ is the drag force per unit dust mass that is expressed as follows:

$$F_D = \frac{18\mu}{\rho_p d_p^2} \frac{C_D Re}{24}$$

where, μ is the molecular viscosity of gas; C_D is the drag coefficient of the dust.

The geometry and the mesh system of the shaft furnace with AGD beams are shown in Fig. 12-2. A uniform velocity for the gas was given at the inlet based on the blast volume. The dust particles with the same properties as the actual production (sizes ranged from 10~90μm, with a density $\rho = 3800 kg/m^3$) were injected into the shaft furnace from the inlet. It was assumed that the particles were reflected when they came in contact with the walls, escaped once reaching the top outlet, and exited the system once they enter the bottom of the domain.

Fig. 12-2 The furnace geometry and the mesh system

12.3 Results and Discussion

12.3.1 Characteristics of Dust Accumulation

A typical experiment results are shown in Fig. 12-3. The corresponding profile of the dust accumulation region in the bustle pipe is shown in Fig. 12-4. In this case, the blast volume and discharging rate was 65m³/h (standard state) and 5.83r/min. This represented a blast volume 235000m³/h (standard state) and a melting rate 120t/h in the actual furnace operation. The slot located below the gas inlet was numbered 0# and the slot located farthest from gas inlet was number as 20#. The angle between the two adjacent slots was 4.5°. There was a large amount of dust deposited in the packed bed in front of slots 8 ~ 12# at $t=3$ min. As the dust was continuously injected, the voids of the packed bed continued to be filled with the dust. The initial deposition zone was formed. As the growth velocity of the deposition zone became larger than the descending velocity, the dust deposition zone spread upward and plugged the slot. The 8 ~ 12# slots were filled completely by the dust at $t=9$ min. The dust accumulation area was seen on the bottom of the bustle pipe. The accumulation area continued to grow in the bustle pipe as time went on, until the dust particles developed into a mountain shaped area. The peak of the mountain shaped area expanded to the top of the bustle pipe. The left corner of the pile expanded to the distal end. This resulted in 14# and 15# slots becoming blocked by dust at $t=24$ min. Due to the limited capacity of the powder spray in this study, the dust injection stopped at $t=27$ min. This was after the initial clogging position and the evolution of the dust accumulation process was determined.

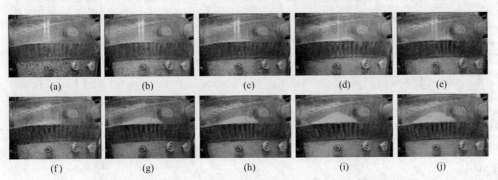

Fig. 12-3 Distribution of powders in bustle pipe zone at blowing rate
65m³/h (standard state), discharging rate 5.83r/min
(a) $t=0$ min; (b) $t=3$ min; (c) $t=6$ min; (d) $t=9$ min; (e) $t=12$ min;
(f) $t=15$ min; (g) $t=18$ min; (h) $t=21$ min; (i) $t=24$ min; (j) $t=27$ min

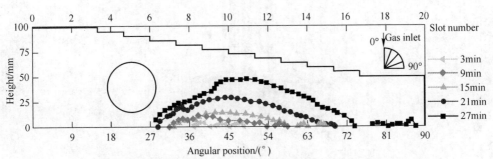

Fig. 12-4 Profile of the dust accumulation at a blowing rate of
65m³/h (standard state), discharge rate of 5.83r/min

The mathematical modeling of the dust movement in the bustle pipe of shaft furnace was used as a good visualization tool. Fig. 12-5 shows the dust distribution in the shaft furnace under a blast volume of 235000m³/h (standard state). The distribution of the dust was discussed in the quarter region of the bustle pipe as the dust was injected into the bustle pipe through the left half of the gas inlet. A majority of the dust entered the shaft furnace through the AGD inlet and followed the gas flow upward. The dust deposits were predominantly distributed in the slots immediately after passing the AGD. The dust distribution is detailed in Fig. 12-6. It can be seen from Fig. 12-6(a) that as the dust diameter increased, the dust deposition rate increased. The dust diameter $d<30\mu m$ had a ratio 4.7% of the deposited dust. When the diameter of $d>50\mu m$, the ratio reached 47.8%. This can be easily explained that larger diameter results in a higher gravity, which could increase the inertia of the downward movement. Thus, in COREX shaft furnace practical production, the larger diameter dust would induce the blockage of the packed bed near slot. The distribution of deposited dust from the various slots is shown in Fig. 12-6(b). The proportion of deposited dust in the various slots was uneven when the AGD beams were added. The ratio in the 8 ~ 12# slots region was higher than that in the other slots. About 57.34% of the deposited dusts appeared in the packed bed in front of 8 ~ 12# slots. This indicated that the choking of the gas slots could have initially started in the 8 ~ 12# slot positions. This phenomenon was consistent with the experimental result seen in Fig. 12-3. The deposited dust concentration in the 8 ~ 12# slots is likely caused by the variation of the gas velocity in the bustle pipe.

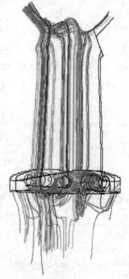

Fig. 12-5 Trajectory of the dust in the shaft furnace

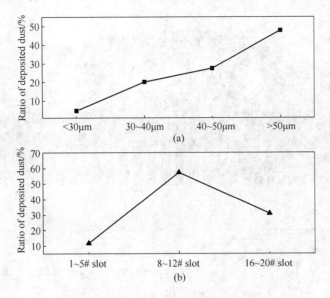

Fig. 12-6 The dust distribution

(a) The proportion of deposited dusts with various sizes; (b) The ratio of deposited dust in various slots

Fig. 12-7(a) depicts the gas vector flow in the bustle pipe. When the AGD was added, the gas vector flow becomes more complicated, particularly around the AGD tube. After the gas passed around the top of the AGD tube, it moved along the tangent of the tube. The main direction of the gas vectors around AGD tube points began at slots 8 ~ 10#, where the initial zone for dust particles accumulation was located. Details of the gas velocity magnitude distribution in the bustle pipe are shown in Fig. 12-7(b). The average velocity in the bustle pipe before AGD was 11.06m/s. After the gas passed through the AGD beams, the gas velocity in bustle pipe sharply decreased. This was due to the nearly 60% gas flow into the furnace through the AGD inlet. The average velocity in the 8 ~ 12# slots region was 3.58m/s. This low velocity resulted from an inability to entrap the dust in the bustle pipe. Some dust escaped from the gas stream and settled on the bottom of bustle pipe, while other dust particles flowed with the gas into the packed bed through the slot. Due to low-velocity gas-dust flow in packed bed, the percolation diffusion capacity of the dust is weak. Further considering the interaction of dust in the voids of packed bed, the low velocity dusts are easier blocked in packed bed in front of the slot.

12.3.2 Effect of Blast Volume

The influence of the blast volume on dust accumulation was initially discussed in our physical simulation. Fig. 12-8 depicts the dust distribution at $t = 2$min under various

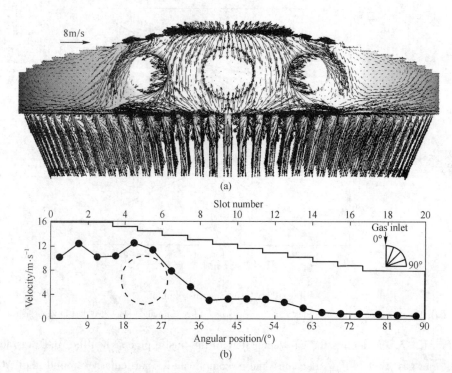

Fig. 12-7 Gas flow in bustle pipe zone
(a) Gas vector flow; (b) Velocity magnitude

blast volumes (standard state). The larger the blast volume was, the further the dust was blown in the bustle pipe. The initial area of the dust in the packed bed increased as the blast volume increased. When the blast volume was large, such as $Q=78 \mathrm{m}^3/\mathrm{h}$, the dust was able to flow into the packed bed through the furthest slot. The wider the dust distribution was, the smaller the dust flux in each slot was when the dust flow rate remained the same. This suggested that the increased blast volume was effective at reducing the dust flux in each slot, while decreasing the clogging effect.

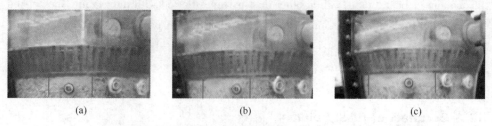

Fig. 12-8 Dust distribution in packed bed at $t=2$min when the discharging rate was 7.29r/min for various blast volumes (standard state)
(a) $Q=52\mathrm{m}^3/\mathrm{h}$; (b) $Q=65\mathrm{m}^3/\mathrm{h}$; (c) $Q=78\mathrm{m}^3/\mathrm{h}$

Fig. 12-9 shows the evolution of the dust accumulation under various blast volumes (standard state). There was no dust accumulation observed in the bustle pipe when the blast volume was 78m³/h (Fig. 12-9(a)). This was caused by the smaller dust flux and the larger gas velocity for the gas-dust flow in the packed bed. The dust was able to penetrate deeper in the voids and the dust deposits were decreased. This would result in reducing the upward spreading speed of the clogging zone, and no dust accumulation would be found in the bustle pipe. As the blast volume decreased, the dust accumulation in the bustle pipe was more likely to occur. The dust accumulation was visible in the bustle pipe when the blast volume was 65m³/h. The initial dust accumulation in the bustle pipe ranged from 8# slot to 12# slot. While comparing with the results under the same blast volume as shown in Fig. 12-4, the dust accumulation in Fig. 12-9 (b) was relatively small. This was caused by a larger discharge rate. The descending velocity of the dust deposition zone in the furnace increased with the increase of the discharging rate. This meant that the deposition zone was more likely to move downwards, rather than to spread into the bustle pipe. The increased descending velocity of burden near the slots was an effective method that lessened the impact of the dust accumulation. However, this method should be controlled under a certain scope to obtain a sufficient gas solid reduction reaction.

A large amount of dust accumulation on the bottom of bustle pipe was found when the blast volume was decreased to 52m³/h. Dust settled on the bottom platform between the slots in the bustle pipe, beginning at 8# slot. Less dust could enter the furnace through the far-end slots. The dust accumulation in bustle pipe was distributed from the 7# slot to the 13# slot at $t = 9$min. This was larger than that in Fig. 12-9(b) at the same time. Dust was continuously injected and the accumulation area continued to grow in the bustle pipe, resulting in more slots becoming blocked. The final peak of the mountain-like dust accumulation area was located near the 10# slot position.

Fig. 12-10 shows the moving track of the dust in the shaft furnace under various blast volumes (standard state). When the blast volume increased from 188000m³/h to 282000m³/h, the percentage of the deposited dust to the injection dust decreased from 33% to 27%. Quantitative analysis of the influence of blast volume (standard state) on the dust deposition rate in different regions is shown in Fig. 12-11. The percent of deposited dust to the injection dust in 8 ~ 12# slots was larger than in the other regions. This indicated that slots 8 ~ 12# were the initial position for dust accumulation. The deposition rate in this region decreased from 19% to 16% when the blast volume increased from

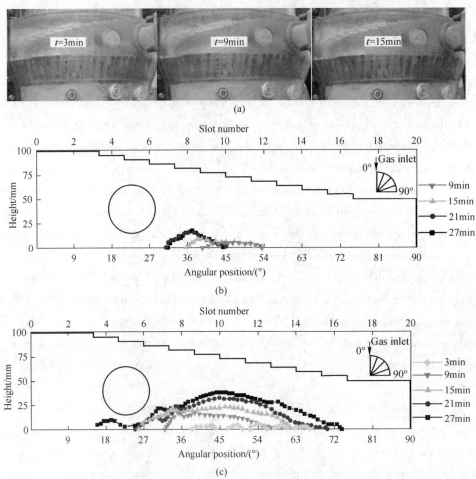

Fig. 12-9 Profile of the accumulation of dust at a discharging rate of 7.29r/min for the various blast volumes (standard state)

(a) $Q=78m^3/h$; (b) $Q=65m^3/h$; (c) $Q=52m^3/h$

188000m^3/h to 282000m^3/h. As the blast volume increased, the gas capacity for the training dust in the bustle pipe increased. The deposition rate in 1~5# slots decreased from 9% to 2%. The deposition rate increased from 5.5% to 10% in the 16~20# slots regions. These results showed a similar tendency that was discussed in our physical simulation, even though the numerical results and the experimental results were obtained from different-sized furnaces.

12.3.3 The Mechanism of Dust Accumulation

The mechanism of dust accumulation in bustle pipe could be described as the installation of AGD that directly affected the gas distribution and the dust motion behavior in the

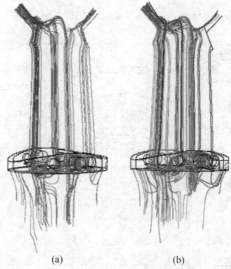

(a)   (b)

Fig. 12-10   Trajectory of the dust under various blast volumes (standard state)

(a) $188000 m^3/h$; (b) $282000 m^3/h$

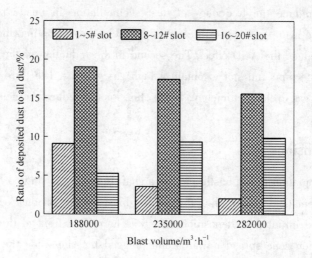

Fig. 12-11   Effect of blast volume (standard state) on dust deposition rate in different regions

bustle pipe. After the gas-dust passed the AGD tube, the gas velocity decreased sharply. The inertial effect resulted in a majority the dust flowing into the packed bed through 8~12# slots. The initial deposition zone was formed in front of 8~12# slots. If the growth velocity of the deposition zone was larger than the descending velocity, the dust deposition zone would spread upward and plug the slot. Dust accumulation in the bustle pipe gradually formed as dust was continuously injected. A schematic diagram of the accumulation of dust in the bustle pipe is shown in Fig. 12-12.

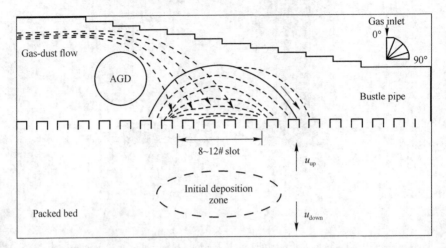

Fig. 12-12 Schematic diagram of the accumulation of dust in the bustle pipe of shaft furnace

In order to reduce the frequency of dust accumulation in bustle pipe, additional efforts should be made in at least three ways. The first lies in optimization the AGD arrangement to reduce the AGD effect. The second is the reduction of the cross-sectional area of the bustle pipe and slot to obtain a high gas velocity. The third is the improvement of the hot gas cyclone efficiency. These efforts could reduce the demand of dust accumulation.

12.4 Conclusions

The dust accumulation in the bustle pipe of a shaft furnace was studied by both a physical and a mathematical model. The effect of the blast volume on dust accumulation was investigated. The initial dust deposition zone was formed in front of the 8 ~ 12# slots, and the deposition zone spread upwards and plugged the slot when the growth velocity became larger than the descending velocity. Dust accumulation in the bustle pipe gradually formed as dust was continuously injected. The initial dust accumulation in the bustle pipe ranged from the 8# slot to the 12# slot. The gas velocity decreased sharply after passing the AGD due to nearly 60% gas flow into furnace through the AGD inlet. This was the main reason for dust accumulation in the 8# to 12# slots region. Increasing the blast volume was effective in reducing the dust flux in each slot, as well as decreasing the clogging effect. The results could be useful for design, control, and optimization of the shaft furnace operation.

References

[1] Zhou Y S. Review of current development of coal-based smelting reduction ironmaking process [J]. Iron Steel, 2005, 40 (11): 1~8.

[2] Lin J J, Song W G, Xia W Y. Progress and improving directions of COREX-3000 in baosteel [J]. 5th Baosteel Biennial Acad Conf SSTLPH, 2013: 499~506.

[3] Wu S L, Du K P, Xu J, et al. Numerical analysis on effect of areal gas distribution pipe on characteristics inside COREX shaft furnace [J]. JOM, 2014, 66 (7): 1265~1276.

13 Numerical Study of Fine Particle Percolation in a Packed Bed

Granular materials are abundant in nature and are one of the most manipulated materials in industry. The transport and mixing of granular materials is a vital operation in various processes in pharmaceutical, agriculture, and chemical industries. During the chute of a mixture of grains of different sizes, small grains will pass through the void between large ones leading to size segregation. A well-known example of size segregation effect is 'Brazil nut segregation'. Generally, when a granular material with two species which have very different diameters, the smaller grains can even drain totally through the larger under gravity, and segregation will occur without any external mechanical action. This is normally termed spontaneous inter-particle percolation. A special industry example can be found in blast furnace ironmaking, where small iron pellet particles are placed upon larger coke particles, and the smaller particles may pass through the larger ones in descending motion under gravity. Besides, during burden charging in ironmaking blast furnace, the mixing and segregation behaviour can be observed, one of the main reasons is particle percolation. Moreover, the chocking of gas slots in bustle pipe of COREX shaft furnace arises mainly from the insufficient percolation of dust particles in the burden. On the other hand, actually, in many industrial processes, granular materials may in cohesive or wet state. This is especially important in ironmaking processes. For example, the significant stickiness between iron ore particles was observed within the temperature range of 600℃ to 675℃. As an attractive force between particles, if the cohesion is significant, substantial difference from the free-flowing behaviour of particulate systems is evident. Hence, in the lump zone of blast furnace or COREX shaft furnace, the cohesion between particles, induced by reduced elastic modulus of relevant materials and in particular, softening-sticking bridge as a result of high temperature operation and reduction reaction of iron ore, may directly affect the percolation behaviour in the packed bed. Thus, it is necessary to understand the fundamentals in the percolation phenomenon in a packed bed, and moreover, explore the physics of cohesive particle percolation for understanding the mixing and segregation of multi-scale burdens in ironmaking processes.

In this chapter, the comparison of sphere and cubical particle on the percolation behavior is studied. The effects of some key variables such as size ratio of packing particles to percolation ones, damping coefficient, and rolling coefficient on percolation behaviors are investigated. This provides important fundamental understandings of percolation of cohesive particles in packed bed and thus beneficial for process understanding and optimization in ironmaking processes.

13.1 DEM Model

Each single particle in a considered system undergoes both translational and rotational motion, described by Newton's 2nd law of motion. The forces and torques considered include those originating from the particle's contact with neighbouring particles, walls and surrounding fluids. The governing equations for translation and rotational motions of particle i with radius R_i, mass m_i and moment of inertia I_i can be written as

$$m_i \frac{dv_i}{dt} = \sum_{j=1}^{k_i} (F_{cn,ij} + F_{dn,ij} + F_{ct,ij} + F_{dt,ij} + f_{e,ij}) + m_i g \quad (13-1)$$

$$I_i \frac{d\omega_i}{dt} = \sum_{j=1}^{k_i} (T_{ij} + M_{ij}) \quad (13-2)$$

where, m_i, I_i, v_i, and ω_i represent mass, rotational inertia, translational velocity, and rotational velocity of particle i, respectively; $F_{cn,ij}$, $F_{dn,ij}$, $F_{ct,ij}$, $F_{dt,ij}$, $f_{e,ij}$, T_{ij}, and M_{ij} represent normal contact force, normal damping force, tangential contact force, tangential damping force, cohesive force, and tangential and rolling friction torque of particle j acting on particle i, respectively; g is the gravitational acceleration; k_i is the number of particles in contact with particle i; and t is time.

Cohesive force between particles can originate from several sources including van der Waals force, electrostatic force, and liquid bridges (capillary forces). This work is focused on a general understanding of percolation behaviour of cohesive particle in a packed bed, thus a simplified model is employed to reduce computational requirement while reasonable and general results can be obtained. The cohesive force $f_{e,ij}$ is expressed as a multiple of the weight of a fine particle, i. e.

$$f_{e,ij} = K\rho g \pi d^3/6 \quad (13-3)$$

Note that the interparticle force expressed in Eq. (13-3) is artificially imposed. It is expressed as a multiple (K) of the weight of a single percolating particle. We assume that the cohesion force only exists when the gap between two particles is less than a critical value (1% of particle diameter), and the cohesion force disappears as soon as the gap exceeds the critical value. This force acts both between percolating-packed particles and percolating-percolating particles. The magnitude of the interparticle force can be easily varied by varying the value of K in Eq. (13-3). A similar approach has been used in previous works for simplification.

For the cubic particles, due to the geometrical non-uniform characteristics, their contact detection is much more complicated than that of spherical particles. Generally, a number of methods have been proposed in the last decade for non-spherical particles representation and contact detection. Among them, the multi-sphere model, introduced by Favier et al. and Abbaspour-Fard, is widely adopted, where the contact determination of the non-spherical particles is easily transformed to the determination of their element-spheres. Moreover, multi-sphere model is simple and flexible, and thus can approximate particles of any shape by the combination of spheres. Therefore, the multi-sphere model is adopted to construct the cuboid particles in the present work.

In multi-sphere model, the total mass m_i of a particle i is given by:

$$m_i = \sum_{s=1}^{n_s} m_{is} - \sum_{j=1}^{n_j} m_{\text{overlap},j} \qquad (13-4)$$

where n_s, n_j, m_{is}, and $m_{\text{overlap},j}$ are respectively the total number of element-spheres, the caps pairs, the mass of sphere s, and the mass of caps pairs in particle i.

In the multi-sphere model, the force F_i of each cuboid particle i is the sum of contact forces on its element-spheres, given by:

$$F_i = \sum_{s=1}^{n_s} F_{is} \qquad (13-5)$$

where F_{is} is the force on element-sphere s in particle p, which is the sum of normal force F_{is}^n and tangential force F_{is}^t acting on the contact points:

$$F_{is} = F_{is}^n + F_{is}^t = \sum_{c=1}^{cs} (F_{isc}^n + F_{isc}^t) \qquad (13-6)$$

where cs is the total number of the contact points on each element-sphere; the normal force F_{isc}^n and tangential force F_{isc}^t are calculated based on Hertz model and the work of Mindlin and Deresiewicz. For the energy dissipation in contact, in addition to the Coulomb friction, the nonlinear viscous damping proposed by Tsuji is used:

$$F_{isc}^n = F_{isc}^{ne} + F_{isc}^{nd} = -\frac{3}{4}E^*\sqrt{R^*}\delta_n^{\frac{3}{2}}n - 2\sqrt{\frac{5}{6}}\beta\sqrt{s_n m^*}\,v_n^{\text{rel}} \qquad (13-7)$$

$$F_{isc}^t = F_{isc}^{te} + F_{isc}^{td} = -S_t\delta_t t - 2\sqrt{\frac{6}{5}}\beta\sqrt{S_t m^*}\,v_t^{\text{rel}} \qquad (13-8)$$

where F_{isc}^{ne} and F_{isc}^{nd} are the normal elastic force and damping force; F_{isc}^{te} and F_{isc}^{td} are the tangential elastic force and damping force; δ_n and δ_t are the normal and tangential overlap; n and t are the unit vector in the normal and tangential directions; S_t is the tangential stiffness, $S_t = 8G^* \sqrt{R^* |\delta_t|}$; v_n^{rel} and v_t^{rel} are the relative normal velocity and tangential velocity; R^* and m^* are the equivalent radius and mass; E^* is the equivalent Young's modulus.

Torque T_{tis}, created by the tangential force on each element-sphere, is given by:

$$T_{tis} = \sum_{c=1}^{cs} r_{isc} \times F_{isc}^t \qquad (13-9)$$

where r_{isc} is the vector between the contact point c and the center of element-sphere s of particle i; and T_{nis} which is created by normal forces when the normal force on the element-sphere does not pass through the center of the particle, is given by:

$$T_{nis} = d_{is} \times F_{is}^n \qquad (13-10)$$

where d_{is} represents the relative position vector between the centroid of particle i and the center of the element-sphere s, and the torque T_i is the sum of the following three parts:

$$T_i = \sum_{s=1}^{ns} (T_{tis} + T_{nis} + T_{ris}) \qquad (13-11)$$

where T_{ris} is the torque generated by the rolling friction, $T_{ris} = -\mu_r |F_{n,ij}| \frac{\omega_{t,ij}}{|\omega_{t,ij}|}$.

13.2 Simulation Conditions

As shown in Fig. 13-1, the simulation setup is made of a cylindrical container of $\phi15D \times 15D$ filled with a packing of monosize large particles of diameter D (which are referred to as packing particles here). The cubical particle is composed of 64 uniform element-spheres with the diameter of d. Note that the shape of the cubic particle constructed by element-spheres is not fully the same as a real cube. The corners of the cubic particle are rounded, and its faces are not smooth. To evaluate the shape similarity between the constructed cubical particle and the perfect cube, they are compared in terms of sphericity. For a perfect cube, the sphericity is 0.806. The surface area and volume of the constructed cubical particle are $32.164d^2$ and $13.288d^3$, respectively. Then, the equivalent diameter (d_e) and sphericity of the cubical particle are obtained as $2.939d$ and 0.843, respectively. The deviation between the constructed cubical particle and the perfect cube is smaller than 5%, indicating the shape similarity between them is acceptable. The simulation domain is a cylindrical container of $\phi15D \times 15D$ filled with a packing of monosize large sphere particles of diameter D. The packing bed is built by random gravity deposition of the particles. This procedure gives a reproducible porosity around 0.4. The ratio of the larger sphere particle diameter and the cubical particle equivalent diameter is 10 ($D = 10d_e$). This size ratio is larger than the geometrical trapping threshold: $\xi_c = \left(\dfrac{2}{\sqrt{3}}-1\right)^{-1} = 6.464$. After these large sphere particles are settled down under gravity to form a stable packing structure, the small cubical particles (percolating particle) are charged on the top of the packed bed. They are generated randomly at the centreline of the column in a circle of diameter of $1D$. These percolating particles pass through the packed bed towards the bottom of the column under gravity. Their dynamic details are recorded for analysis. The parameters used in the present simulations are listed in Table 14-1. In this work, the same simulation process is repeated three times and each packing is rebuilt for each simulation. Each numerical data which are presented in this paper are collected from a statistical mean of several simulations. The error bars are deduced from these replicate simulations.

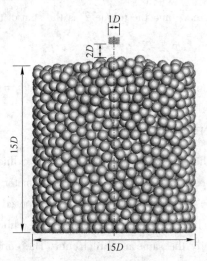

Fig. 13-1 Geometry of the model used in this work

13.3 Results and Discussion

13.3.1 Comparison between Cubical and Sphere Particles

When a smaller percolating particle is put on a packed bed of larger particles, it may move down through the bed in longitudinal and transverse directions under gravity and interactions with the packing particles. The percolation velocity reflects, to a certain degree, the dispersion property of percolating particles. Therefore, the percolation velocity and dispersion are two important percolation indicators. The percolation behavior will be compared between cubical and sphere particles in terms of single particle and a blob of particles.

13.3.1.1 Single Particle

Fig. 13-2 shows the evolution of vertical position of a single cubical particle and a single sphere particle in the packed bed, respectively. The time is set to be proportional to the free fall time to pass a single large sphere diameter. The single cubical and sphere particles descend from the same position, and they have the same equivalent diameter. It is found that both the normalized heights almost evenly decrease with the normalized time, which means a constant mean velocity in both cases. The slope of cubical particle is larger than that of sphere particle, and the former has a higher descending velocity.

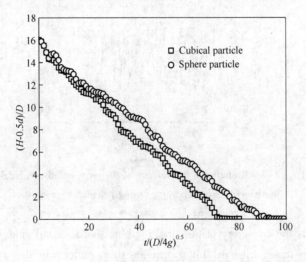

Fig. 13-2 Evolution of vertical position of a single cubical and sphere particle

Although a constant mean velocity of the single cubical particle and sphere particle can be observed, the variation of the velocity with time is very complex. The evolution of the velocity of the single cubical particle and sphere particle is shown in Fig. 13-3. The velocity is in units of free fall velocity reached after falling over one packed particle. It can be seen that the velocity magnitudes of the single cubical particle and sphere particle fluctuate largely and irregularly. Especially, the fluctuation of the x/y-velocity (horizontal direction) denotes the stochastic motion of the single particle. Such stochastic motion is the main reason for the percolation particle dispersing in a packed bed. The fluctuation of the z-velocity (vertical direction) is caused by the impact between the percolating and packing particles. The amplitude of the cubic particle is smaller than that of sphere particle, which indicates the resistance of the downward movement of cubical particle is less. The main reasons will be discussed later.

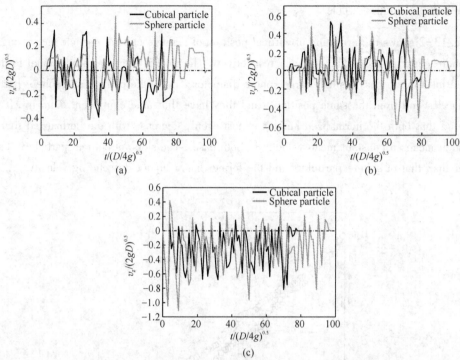

Fig. 13-3 Evolution of the velocity of the representative cubical particle and sphere particle during percolation process
(a) x-direction; (b) y-direction; (c) z-direction

Fig. 13-4 plots the trajectory of the single cubical particle and sphere particle in the packed bed. Both trajectories exhibit a tortuous drop pattern. In the upper part of the packed bed, the trajectories of the two particles are almost the same. However, with the downward movement, the deviation of the trajectories caused by colliding with the packed particles becomes larger, and the two type particles descend along different void channels. At the exit of the packed bed, the transverse dispersion distance of sphere particle is farther than that of cubical particle. This is expected that the descending velocity of sphere particle is smaller, and the particle has a longer residence time in the packed bed. Thus, the sphere particle has more chance to collide with the packed particles, which is benefit for exploring laterally the porous.

13.3.1.2 A Blob of Particles

The evolution of normalized mean vertical velocity of particle group is presented in Fig. 13-5. It can be seen that the mean velocity of cubical and sphere particles shows the similar features. At the very beginning, both cubical and sphere particles free fall to

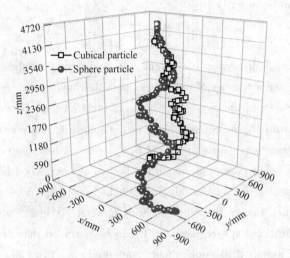

Fig. 13-4 Trajectory of the single cubical particle and sphere particle in packed bed

the packed bed, and the velocities increase rapidly. After a short time of free fall, the mean velocities decrease progressively towards a stead value. These phenomena are related the previous works: for sphere particles falling down in a random porous media, the vertical velocity is a constant. The cubical particles show the same variation trend as the sphere particles, expect the mean vertical velocity larger than that of sphere particles.

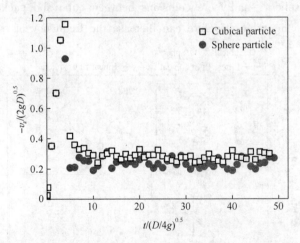

Fig. 13-5 Evolution of mean vertical velocity during percolation process

The radial dispersion of percolating particles is a random walk process. A theoretical model has been proposed by Bridgwater et al to describe the transverse dispersion of sphere percolating particles. For a large packed bed with the wall effect ignored, the radial dispersion can be described by

$$\ln\left(\frac{N_0}{N_0 - N}\right) = \frac{1}{4E_r t} r^2 \qquad (13\text{-}12)$$

where E_r is the radial dispersion coefficient, N_0 is the total number of the percolating particles, and N is the number of percolating particles having centers within radius r at time t. As can be seen in the Eq. (13-12), the relationships between $\ln[N_0/(N_0-N)]$ and r^2 relies on the percolating time. In the following analysis of dispersion, we will only consider one percolating time, $t/(D/4g) = 47$, where the percolating particles disperse sufficiently and no particle reach the bottom of the packed bed.

Fig. 13-6 illustrates the relationship between $\ln[N_0/(N_0-N)]$ and r^2 for cubical and sphere particles. Linear fits obtained with unweighted least-squares method are presented for cubical and sphere particles. It can be observed that the cubical particles also satisfy the radial dispersion theory proposed by Bridgwater. The correlation coefficient of the cubical particle fitted line is 0.991, which is smaller than that of sphere particles. Compared with the sphere particles, the slop of the fitted line of cubical particle is larger, and the radial dispersion coefficient is smaller. The cubical particles have a less off-center of the redial dispersion distance. This is because of the shape of the percolating particle. Although the percolating cubical and sphere particles have the same equivalent diameters, the diagonal length of the cube is 1.47 times the diameter of the sphere, and more collisions between cube and packed particles will occur. Moreover, the cubic structure can increase the frequency of eccentric collision

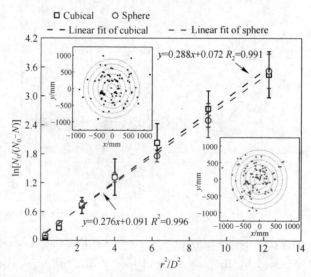

Fig. 13-6 Plot of $\ln[N_0/(N_0-N)]$ against r^2 for cubical and sphere particles

between cube and packed particles, resulting in a larger rotational kinetic energy than that of sphere, as shown in Fig. 13-7(a). Thus, the cubical particles have a smaller probability of moving outside a pore in the horizontal and cause a greater probability of passing through the vertical pore. This is also the main reason that the cubical particles have a larger mean vertical velocity. The translational kinetic energy of cubic and sphere particles shown in Fig. 13-7(b) also indicated that the cubic particles have a larger value. The average translational kinetic energy of cubic particles is 20.26×10^{-3} J, which is 3.6% higher than that of sphere particles.

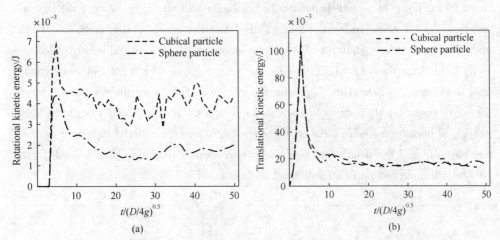

Fig. 13-7 Comparison of rotational kinetic energy and translational kinetic energy between cubical and sphere particle
(a) Rotational kinetic energy; (b) Translational kinetic energy

13.3.2 Effect of Cohesive Force on Percolation Behavior

When a smaller percolating particle is put on a packed bed of larger particles, it may move down through the bed in longitudinal and transverse directions under gravity and interactions with the packing particles. The longitudinal direction is referred to the flow direction, and the transverse one is referred to the direction perpendicular to the flow direction. The percolation velocity reflects, to a degree, the dispersion property of percolating particles. Therefore, the percolation velocity and dispersion are two important percolation indicators.

13.3.2.1 Percolation Velocity

Fig. 13-8 shows the evolution of normalized mean height and normalized mean vertical velocity of percolating particles under different cohesive force conditions when $D/d = 8$,

$c = 0.3$ and $\mu_r = 0.001D$. The time is set to be proportional to the free fall time to past a single large sphere diameter and the velocity is in units of free fall velocity reached after falling over one large particle. It is indicated that the normalized height and mean velocity decreases rapidly once percolating particles collide to the packed particles. Then the percolating particles would move toward to the opening of the orifice and percolate among packed particles. For those percolating particles with cohesionless force, the normalized height decrease gradually and the mean velocity decreases progressively towards a steady value. This phenomenon can be related to previous results: for the case of particles falling down in a random packed bed of larger particles, the vertical velocity is a constant. For the cohesive force $f_e = 2mg$, the variation trends of the normalized height and mean velocity are similar to that of cohesionless case, except the mean vertical velocity larger than that of $f_e = 0mg$. The reason will be discussed in next section by combining the dispersion behaviour. While for the case with $f_e = 8mg$, the normalized height decreases progressively towards to a steady value and the mean velocity is reduced to zero. This is because cohesive force is strong enough to resist the inertia motion of percolating particles and the insufficient percolation can occur. Percolating particles will finally adhere to the packed particles and blockage can be observed in this condition.

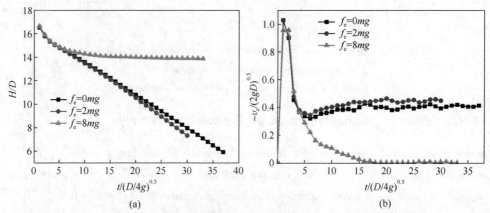

Fig. 13-8 Evolution of height and velocity of percolating particles for different cohesive forces when $D/d = 8$, $c = 0.3$ and $\mu_r = 0.001D$

(a) Height; (b) Velocity

Fig. 13-9 presents the statistic distributions of residence time for different cohesive forces when $D/d = 8$, $c = 0.3$ and $\mu_r = 0.001D$. As the percolating particles directly stick on the surface of packed particles when the cohesive force $f_e = 8mg$, the residence time for this condition is not considered. It has been obtained in the previous experimental studies that the residence time distribution is roughly similar to a normal distribution. Similar trend can also be observed in the present simulation. Besides, with the cohesive force increasing from $0mg$ to $2mg$, the distribution curve shifts to the left. It is because the larger the cohesive force, the higher the particle percolation velocity, and then the less time for the particles to reach the bottom of the packed bed.

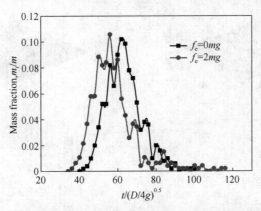

Fig. 13-9 Statistic distributions of residence time with different cohesive forces for $D/d=8$, $c=0.3$ and $\mu_r=0.001D$

13.3.2.2 Dispersion Behavior

The dispersion of percolating particles is a random walk process. The percolation flow of the small particle is subject to stochastic motion, which is caused by the interaction between percolating and packing particles. Such stochastic motion leads to dispersion of percolating particle within a region in the packed bed. Fig. 13-10 illustrates the distribution function of particle positions at the exit of packed bed for different cohesive forces. It can be seen that the smaller the cohesive force, the more off-centre the dispersion distance. Especially for the case $f_e=8mg$, the insufficient percolation occurs and the percolating particles almost stick in the central part of the packed bed. In order to quantitatively describe the dispersion behaviour, detailed information, such as the transverse and longitudinal dispersion coefficients, will be discussed.

The DEM model is possible to access individual particle positions, anywhere at any time, inside the packing of larger spheres. The position of particle k in the horizontal plane can be described as $r_k^2=x_k^2+y_k^2$, where x_k and y_k are the particle positions. So, the variance of the position distributions of the N moving particles in this plane is

$$\langle(\Delta r)^2\rangle = \frac{1}{N}\sum_{k=1}^{N}(r_k-\langle r\rangle)^2 \qquad (13-13)$$

where $\langle r\rangle = \frac{1}{N}\sum_{k=1}^{N}r_k$. In the same manner, if the particle position in the flow direction

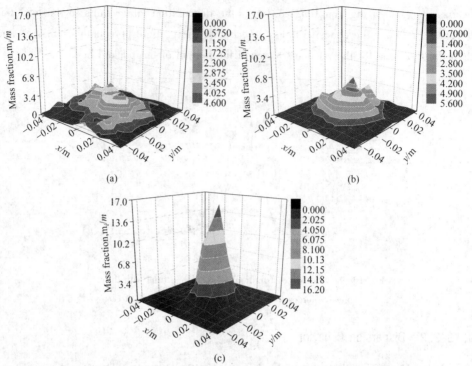

Fig. 13-10 Distribution function of particle positions at the exit of packed
bed for cohesive force when $D/d=8$, $c=0.3$ and $\mu_r=0.001D$

(a) $f_e=0mg$; (b) $f_e=2mg$; (c) $f_e=8mg$

is denoted by z_k, it can be written as

$$\langle(\Delta z)^2\rangle = \frac{1}{N}\sum_{k=1}^{N}(z_k - \langle z\rangle)^2 \tag{13-14}$$

where, $\langle z \rangle = \frac{1}{N}\sum_{k=1}^{N} z_k$.

Then, the transverse and axial dispersion coefficients $D_\perp$ and $D_\parallel$ can be defined from the time evolution of $\langle(\Delta r)^2\rangle$ and $\langle(\Delta z)^2\rangle$ with Einstein-Smoluchowski equation as follows.

$$\langle(\Delta r)^2\rangle = 2D_\perp t \text{ and } \langle(\Delta z)^2\rangle = 2D_\parallel t \tag{13-15}$$

Fig. 13-11 presents the variances, $\langle(\Delta r)^2\rangle$ and $\langle(\Delta z)^2\rangle$ calculated with Eqs. (13-13) and (13-14), of particle position distribution for different cohesive forces versus time. Two main features can be observed from Fig. 13-11. The first is that a transition state occurs in the initial period. When particles are launched on top of the packed bed, a quantity of them rebounces on the upper packing surface. Moreover, due to the

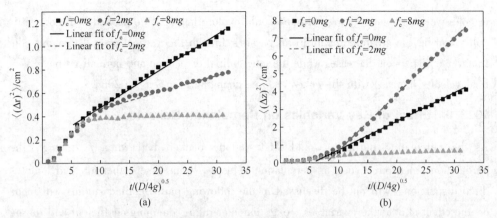

Fig. 13-11 Time evolutions of different cohesive forces when $D/d=8$, $c=0.3$ and $\mu_r=0.001D$
(a) $\langle(\Delta r)^2\rangle$; (b) $\langle(\Delta z)^2\rangle$

presence of other small particles, some particles cannot enter in the porous space since pores are already filled with other particles. These effects can lead to a penetration delay. Thus, the percolating particles need time to reach a diffusive stage. The more probable value of the transition time is close to a value of 5.7 $(D/4g)^{1/2}$ when $f_e=0$. This value is corresponds to the free fall time over a distance of $2D$, integrating the distance between the two initial packing position and the first grain thickness. With increasing the cohesive force, the transition time slightly increases. The other feature is the linear evolution of $\langle(\Delta r)^2\rangle$ and $\langle(\Delta z)^2\rangle$ are observed after the transition state. Linear fits obtained with unweighted least-squares method are presented for cohesive force $f_e=0mg$ and $f_e=2mg$. We can notice that the linear evolution of the two variances with time is a typical proof of a diffusive property. It should be pointed out that for the larger values of t, the deviation from the fits in Fig. 13-11 can be explained by the finite size of our simulation and some percolating particles have already reached the bottom of the packed bed. Diffusive motion of an isolated particle and a blob of particles were found by Ippolito et al and Lomine et al, respectively. Our simulations prove the same behaviour for smaller cohesive force. However, as mentioned above, the insufficient percolation occurs for the case $f_e=8mg$, and no diffusive motion can be observed. From the Fig. 13-11, it also can be seen that the transverse dispersion of $f_e=2mg$ is smaller than that of $f_e=0mg$, while the longitudinal dispersion becomes larger when cohesive force changes from $0mg$ to $2mg$. The main reason can be explained as: When percolating particles meet packing particles, they would move downwards and experience multiple collisions. All the contact between the percolating and packing particles would directly decrease the bouncing due to the cohesive force acting as an attractive force. Under the effect of cohesive force, percolating particles would explore laterally the packed bed more difficultly and have a greater probability of passing through the vertical pore. Therefore, the transverse dispersion decreases while the longitudinal dispersion and normalized mean vertical velocity increase with the cohesive force changing from $0mg$ to $2mg$.

13.3.3 Effect of Key Variables on Percolation Behavior

As the percolating particles would stick in the packed bed when $f_e = 8mg$, the percolation behaviour, such as percolation velocity, transverse dispersion and longitudinal dispersion, will not be discussed in the following part. Detailed studies will focus on the effect of other key variables, e. g. diameter ratio, damping coefficient and rolling coefficient, on percolation behaviour. In this section, the fitted lines obtained by the unweighted least-squares method are represented in the main diffusion stage, and all the dispersion coefficients are determined by the slope of the fitted lines.

13.3.3.1 Diameter Ratio D/d

Fig. 13-12 shows the variation of percolation velocity with diameter ratio of packing particle to percolating particle for different cohesive force. It can be seen that with increasing the particle diameter ratio, the percolation velocity increases. This can be easily explained by considering a single particle falling down toward to a pore. The particle trajectory must be aligned with the pore hole to pass through it without bouncing around for big ratio of particle size. If not, the smaller the size ratio is, the more important is the collisions of particles. In such a case, the downward motion of the percolating particle is more difficult and leads to a decrease in the percolation velocity.

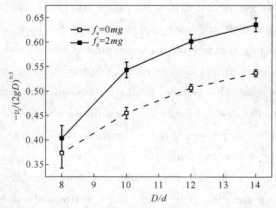

Fig. 13-12 Variation of percolation velocity with particle size ratio for $c=0.3$ and $\mu_r=0.001D$

Fig. 13-13 shows the variations of transverse dispersion and longitudinal dispersion coefficients with the particle diameter ratio. Fig. 13-13(a) demonstrates that the slope of the fitted line increases with the diameter ratio. With the diameter ratio D/d increases from 8 to 14, the transverse dispersion coefficient of $f_e = 0mg$ and $f_e = 2mg$ increases from 0.785 to 1.21cm^2/s and 0.476 to 0.902cm^2/s, respectively, as shown in the up left region of Fig. 13-13(a). The smaller the percolating particles are, the longer distance they can laterally move. On the other hand, Fig. 13-13(b) shows that, the longitudinal dispersion $D_{\parallel}$ decreases with the increasing of the diameter ratio. The longitu-

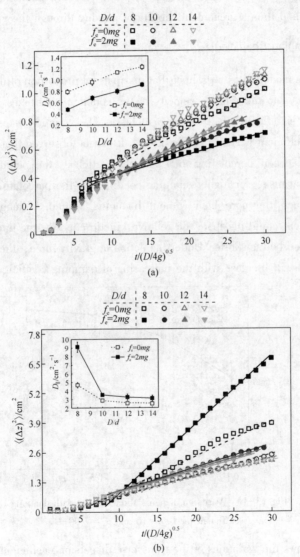

Fig. 13-13 Variations of transverse dispersion and longitudinal dispersion coefficients with D/d
(a) Transverse dispersion; (b) Longitudinal dispersion

dinal dispersion coefficient of $f_e=0mg$ and $f_e=2mg$ decreases from 4.674 to 2.525 cm²/s and 9.12 to 3.105 cm²/s respectively, with increasing the diameter ratio. The transverse dispersion coefficient varies in opposition to the longitudinal dispersion. For the smaller diameter ratio, the pore throats acting like gates in packed bed would create a "gate or valve effect" which could lead to an impedance on the transverse motion of percolating particle, hence reduces the transverse dispersion. On the other hand, when D/d is small, the particles have a smaller probability of moving outside a pore in the horizontal direction and thus a greater probability of passing through the vertical pore.

13.3.3.2 Damping Coefficient

Damping force is modelled as a dashpot that dissipates a proportion of the relative kinetic energy. It is related to the relative velocity of the contacting particles. So, the damping coefficient has a significant effect on the percolation. The variation of percolation velocity with damping coefficient is shown in Fig. 13-14. It can be observed that for higher damping coefficient between percolating and packing particles, the percolation velocity is higher. This is because, for higher damping coefficient, the percolating particles would easily pass through the pore hole without bouncing around. Although increasing the damping coefficient could dissipate the relative kinetic energy, the effective percolation path in the packed bed is shorter. Under the action of gravity force, the mean velocity in vertical direction will increase with the increasing of damping coefficient.

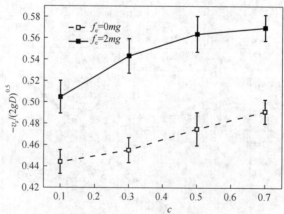

Fig. 13-14 Variation of percolation velocity with damping coefficient for $D/d=10$ and $\mu_r=0.001D$

Fig. 13-15 shows the variations of the transverse dispersion coefficient and longitudinal dispersion coefficient with damping coefficient. Fig. 13-15(a) reveals that the transverse dispersion coefficient decreases with the damping coefficient. The smaller the damping coefficient is, the longer distance they can laterally move. From the Fig. 13-15(b), it can be seen that the longitudinal dispersion $D_\parallel$ increases with the increasing of the damping coefficient. For example, when the cohesive force $f_e = 2mg$, with increasing of damping coefficient, the transverse dispersion coefficient decreases from 0.853 to 0.542 cm²/s while the longitudinal dispersion increases from 3.041 to 4.095 cm²/s. The main reason can be explained as: damping coefficient is related to restitution coefficient. For a fixed Young's modulus, increasing damping coefficient decreases restitution coefficient. Thus,

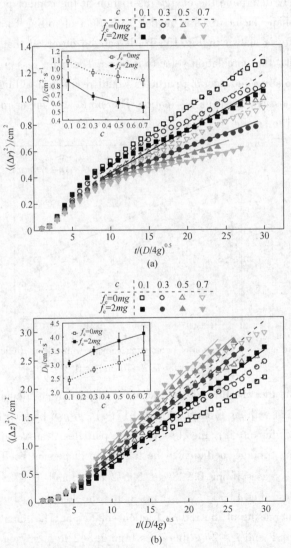

Fig. 13-15 Variations of transverse dispersion and longitudinal dispersion coefficients with c
(a) Transverse dispersion; (b) Longitudinal dispersion

with a smaller restitution coefficient, the percolating particles bounce less and can more difficult find their way into the radial space between packing particles, giving a decrease in transverse dispersion coefficient. This way, longitudinal crossing of individual pore is easier and leads to an increase in the longitudinal dispersion coefficient.

13.3.3.3 Rolling Friction Coefficient

Rolling friction provides an effective mechanism to control the translational and rotational motions and largely determine the energy dissipation at the contact point. It can also be considered as a shape factor in DEM model. The effect of rolling friction coefficient on percolation velocity is shown in Fig. 13-16. It can be observed that for higher rolling friction coefficient, the percolation velocity is lower. The main reason is that the contact between the percolating and packing particles would result in a rolling resistance due to elastic hysteresis losses or viscous dissipation. Therefore, the larger rolling friction coefficient would lead to smaller percolation velocity.

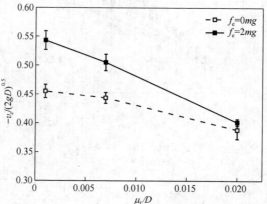

Fig. 13-16 Variation of percolation velocity with rolling friction coefficient for $D/d=10$ and $c=0.3$

Fig. 13-17 shows the variations of the transverse dispersion coefficient and longitudinal dispersion coefficient with μ_r. From the Fig. 13-17(a), it can be seen that the larger the rolling friction coefficient is, the less is the particles transverse dispersion. With increasing of rolling friction coefficient, the transverse dispersion coefficient of $f_e=0mg$ and $f_e=2mg$ decreases from 0.948 to 0.672 cm^2/s and 0.675 to 0.356 cm^2/s, respectively. Besides, Fig. 13-17(b) shows that, the longitudinal dispersion $D_\parallel$ increases with the increasing of the rolling friction coefficient. The longitudinal dispersion coefficient of $f_e=0mg$ and $f_e=2mg$ increases from 2.842 to 5.337 cm^2/s and 3.502 to 11.749 cm^2/s, respectively. The main is that, for smaller rolling friction coefficient, the energy dissipation is reduced and the particles can explore laterally the porous medium more easily due to the more chance to bounce around the packing particles. This observation is consistent with the results of the study of Yu and Saxén, where a low rolling friction between pellets promotes the percolation in ironmaking blast furnace. On the other hand, when the rolling friction coefficient is increased, more relative kinetic energy can be dissipated, and the gravity force is more important. Hence, the longitudinal crossing of individual pore is easier and leads to an increase in the longitudinal dispersion coefficient.

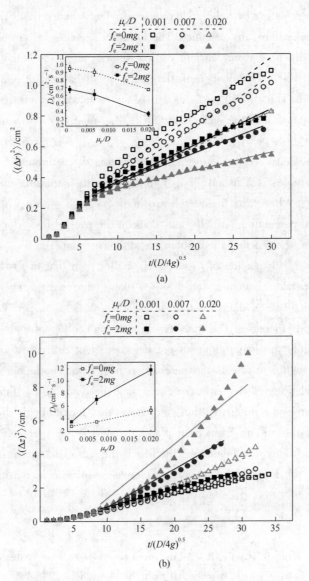

Fig. 13-17 Variations of transverse dispersion and longitudinal dispersion coefficients with μ_r
(a) Transeverse dispersion; (b) Longitudinal dispersion

13.4 Summary

The percolation behavior plays an important role in many pyrometallurgy processes. The key phenomena including percolation velocity, residence time distribution, longitudinal and transverse dispersion, of fine particles in a packed bed is studied numerically by means of DEM approach. The effects of key variables such as the diameter ratio of packing particle to percolating particle, damping coefficient, and rolling friction coefficient on percolation behaviours are examined. The following conclusions are obtained:

The cubical particles have a constant vertical velocity along the height during percolating in the packed bed. The mean vertical velocity of cubical particles is larger than that of sphere particles. The cubical particles satisfy the radial dispersion theory proposed by Bridgwater. Compared with the sphere particles, the cube has a less off-center of the radial dispersion distance.

The vertical velocity of percolating particles moving down through a random packed bed of larger particles is constant, but it increases with increasing the cohesive force from $0mg$ to $2mg$. While for a higher cohesive force, e. g. $f_e = 8mg$, insufficient percolation occurs and percolating particles may stick in the packed bed. The percolating particles moving in the packed bed shows a diffusive property for smaller cohesive force. The transverse dispersion of $f_e = 2mg$ is smaller than that of $f_e = 0mg$, while the longitudinal dispersion becomes larger when cohesive force changes from $0mg$ to $2mg$. With the increase of diameter ratio of packing particles to percolating ones, the percolation velocity increases. The transverse dispersion coefficient increases with the diameter ratio, while the longitudinal dispersion coefficient decreases with the diameter ratio. Damping coefficient affects the percolation behaviour. For higher damping coefficient, the percolation velocity is larger. Increasing the damping coefficient reduces the transverse dispersion coefficient, but increases the longitudinal dispersion coefficient. Increasing the rolling friction coefficient reduces the percolation velocity. The larger the rolling friction coefficient is, the less is the particles transverse dispersion. On the other hand, the longitudinal dispersion increases with the increasing of the rolling friction coefficient.

References

[1] Metzger M J, Remy B, Glasser B J. All the Brazil nuts are not on top: Vibration induced granular size segregation of binary, ternary and multi-sized mixtures [J]. Powder Technol, 2011, 205: 42~51.

[2] Yu Y W, Saxén H. Effect of DEM parameters on the simulated inter-particle percolation of pellets

into coke during burden descent in the blast furnace [J]. ISIJ Int. , 2012, 52: 788~796.
[3] Remond S. DEM simulation of small particles clogging in the packing of large beads [J]. Physica A, 2010, 389: 4485~4496.
[4] Scott A M, Bridgwater J. Self-diffusion of spherical particles in a simple shear apparatus [J]. Powder Technol, 1976, 14: 177~183.
[5] Lomine F, Oger L. Dispersion of particles by spontaneous interparticle percolation through unconsolidated porous media [J]. Phys. Rev. E, 2009, 79: 051307.
[6] Fan H J, Mei D F, Tian F G, et al. DEM simulation of different particle ejection mechanisms in a fluidized bed with and without cohesive interparticle forces [J]. Powder Technol, 2016, 288: 228~240.
[7] Mindlin R D, Deresiewicz H. Elastic spheres in contact under varying oblique forces [J]. J. Appl. Mech. Trans. ASME, 1953, 20: 327~344.
[8] Tsuji Y, Tanaka T, Ishida T. Lagrangian numerical simulation of plug flow of cohesionless particles in a horizontal pipe [J]. Powder Technol, 1992, 71: 239~250.

14 Dynamic Analysis of Blockage Behavior of Fine Particles in a Packed Bed

Blockage behavior of fine particles in a packed bed of large particles may happen in many industrial processes. The blockage of small particles in pores can modify the permeability of the crossed medium and the flow of percolating particles. One special industry example can be found in COREX melter gasifier, where the generated unburnt pulverized coal flow through the packed bed in the furnace and some parts of these fine particles accumulate in the pores of the packed bed. The blockage of the fine particles decrease the voidage of the packed bed and narrow the flow paths among the packed particles. Consequently the permeability of the packed bed is deteriorated, and the extreme powder accumulation causes the malfunction of the blast furnace. Recently, the chocking of slot in COREX shaft furnace is also found to be closely related to the fine particles accumulation in the packed bed near the gas slot. Thus, it is important to understand the blockage behavior involving the mechanisms of fine particles in a packed bed of ironmaking reactors.

In this chapter, DEM simulation is used to explore the blockage features of fine particles through an unconsolidated porous media made with large sphere particles. First, the blockage features between two rolling friction coefficient particles is compared, and the influence of key variables are also examined. In the second part, the influence of cohesive force between particles on blockage behaviour of fine particles is investigated.

14.1 Blockage Behavior of Fine Particles

14.1.1 Simulation Conditions

The simulation domain which is a cylindrical container of $\phi 15D \times 15D$ filled with a packing of monosize large sphere particles of diameter D (the same as showed in Fig. 13-1). The porosity of the packed bed is around 0.4. Previous work indicated that the limit diameter ratio of larger packed particle and fine particle for a fine particle free passaging through void is 6.464. However, it should be pointed, when a large number of fine particles pass through an orifice, the interaction between fine particles is easy to cause blockage phenomenon. In the actual operation of BF with PCI, it demands the particle size of coal should be less than 0.074mm. However, the size distribution of packed coke particles at the borders of raceway is extensive. The diameter ratio of packed coke particles to unburned coal in a local area will be close to 10. Thus, in this work, the ratio of the larger sphere particle diameter and the fine particle diameter is 10 ($D=10d$) which is larger than the geometrical tapping threshold. After these large sphere particles are settled down under gravity to form a stable packing structure, the small fine particles are charged on the top of the packed bed. They are generated randomly at the centreline of the column in a circle of diameter of $1D$. In the practice, the fine particles are entrained by the reducing gas and flows mainly upward. This study focuses on the fundamental blockage behavior of fine particles moving through a packed structure with collisions among the fine particles and with the packed particles. Thus no gas flow in the packed structure is considered and the small fine particles pass the packed structure under gravity. These fine particles pass through the packed bed towards the bottom of the column under gravity. Their dynamic details are recorded for analysis. The parameters used in the present simulations are listed in Table 14-1.

Table 14-1 Parameters used in the simulation

Variables	Value
Diameter of fine particle d/m	0.029239
Diameter of packed particle D/m	0.2939
Shear modulus E/Pa	1×10^7
Poisson's ratio v_p	0.3
Restitution coefficient e	0.6
Sliding frictional coefficient μ_s	0.3
Rolling frictional coefficient μ_r	$0.0068D$ (Base Case), $0.001D$
Initial velocity v	0 (Base case), $(2gD)^{0.5}$, $(4gD)^{0.5}$, $(6gD)^{0.5}$
Percolating particle number N	100, 500 (Base Case), 1000, 1500
Time step Δt/s	1.0×10^{-5}

14.1.2 Blockage Distribution and Mechanism

Fig. 14-1 shows the dispersion of fine particles flow for two conditions. The dots in the figures denote fine particles and the packed-bed particles are not shown. The time is set to be proportional to the free fall time to pass a single large sphere diameter. During the passage process of the fine particles in the packed bed, a part of fine particles is deposited and clogged in the voidage of the bed after colliding with the packed particles and fine particles itself, while the other fine particles percolate the packed bed under the gravity. At $t/(D/4g)^{0.5}=60\sim80$, the fine particles begin to reach the bottom of the packed bed. Comparing the passage behavior of the fine particles, it can be found that the velocity of the fine particles in case (a) is somewhat lower than that in case (b). This is main due to the energy losses or viscous dissipation when the collisions taking place. Thus, the smaller rolling friction coefficient causes larger percolation velocity. At $t/(D/4g)^{0.5}=140$, sufficient time has passed and the number of the fine particles in the packed bed becomes constant. The blockage fine particles not only exist as agglomerations but also occur discretely. The total fine particles clogging ratio in case (a) is 39.4%, which is 3.6% higher than that in case (b).

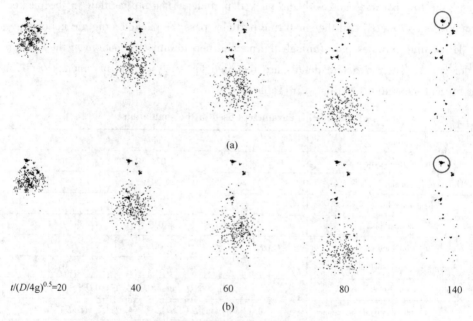

Fig. 14-1 Distribution of fine particles
(a) $\mu_r=0.0068D$; (b) $\mu_r=0.001D$

The distribution of clogging ratio of fine particles along the longitudinal direction is shown in Fig. 14-2(a). It can be seen that the distribution shows a similar tendency under two different rolling friction. More than 70% of the clogging particles are distributed in the area above 2800mm height of the bed. This is mainly because in the process of fine movement in the bed, the more it moves downward, the more dispersed, and the lower mass flow rate of the fine particles in the local voids, which is not conducive to the formation of blockage. Fig. 14-2(b) represents the distribution of clogging ratio of fine particles along the radial direction. According to the multiple of radius $R = 0.5D$, several circles are divided to analyze the radial distribution of clogged fine particles. Since the initial fine particles are located at the centerline of the column in a circle of radius of R, most of the blockage fine particles are deposited in the radial range of less than $3R$. From the distribution of clogging particles, it can be found that when the fine particle just enters the packed bed, the passage and diffusion is not sufficient, which is easy to induce the interaction of fine particles and form agglomeration deposition.

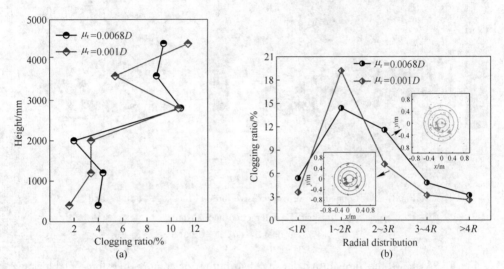

Fig. 14-2　Distribution of clogging particles along the longitudinal and radial direction
(a) Longitudinal direction; (b) Radial direction

Comparison of mechanical behavior at $t/(D/4g)^{0.5} = 140$ in the local cluster that is indicated by red circle in Fig. 14-1 is shown in Fig. 14-3. The bottleneck of void spaces in a packed bed is the key point for cluster formation of clogged fines. Some fine particles simultaneously enter the narrow part of the orifice at the same time and they are stuck on the packed particles. The first accumulation particles in a bottleneck caused by the interaction between the fine and packed-bed particles and fine and fine particles, have the larger force. These fine particles called 'bottleneck particles' in the rest of the paper, are shown in red in Fig. 14-3. After the bottleneck fines are formed, the following fine particles collide to the preceding bottleneck fines and these collision bring little motion of these stuck fine, and finally these fine particles settle there. These fine particles piled on the bottleneck fines have a small force and show blue in Fig. 14-3. Therefore, the amount of fine particles in an orifice directly influence the probability of forming bottleneck particles, thus affecting the fine particles clogging rate. In addition, the increase of fine particles percolation velocity may impact the bottleneck fines formed, and would have a certain impact on the deposition rate. The effects of these two factors on fine particles blockage will be discussed in the next section.

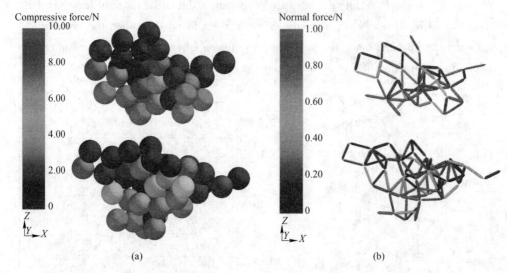

Fig. 14-3 Comparison of compressive force of blockage particles and fine particle interaction normal force
(a) $\mu_r = 0.0068D$; (b) $\mu_r = 0.001D$

Fig. 14-4 shows the probability density distribution of compressive force of blockage particles under different μ_r. It can be observed that the magnitude of compressive force varies in a large range for both cases. Both the profiles show an early-occurred peak and the peak value increases from 18.89% to 23.28% when the rolling friction coefficient decreases from 0.0068D to 0.001D, which indicating that the number of weak particle increases. In the range greater than 10N, the sum of the probability density increases from 15.21% to 23.28% with the increase of friction coefficient, indicating the number of bottleneck particles increases. This phenomenon is consistent with that in Fig. 14-3. With the increase of rolling friction coefficient, the average compressive force increases from 5.269N to 6.35N.

Based on the above simulation results, the blockage mechanisms can be explained as follows. For the discretely clogging powder particles, the mechanical is mainly due to drag force and friction between one small particles rolling very slowly on the surface of large particles whose spacing is close to the diameter of powders. For the cluster, Fig. 14-5 shows the schematic diagram of the blockage of fine particles in packed bed. The blockage is mainly due to mechanical interactions between fine particles,

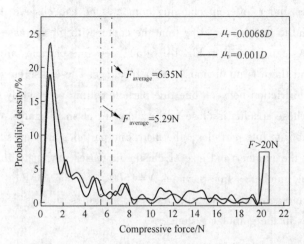

Fig. 14-4 Probability density distribution of compressive force of blockage particles under different μ_r

which can create arches on packed bed and stop the flow. As stated by Civan, the pore throat acts like gate connecting the orifice and creates a 'gate or vale effect', indicated by a severe reduction of permeability as they are plugged by fine particles and shut off. When the fine particles form a bridge across the pore throat of the orifice, the bottleneck of void space becomes the starting point for blockage formation.

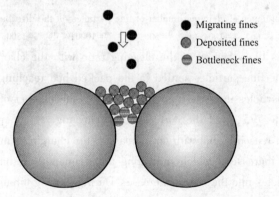

Fig. 14-5 Schematic diagram of fine particles deposition in the packed bed

14.1.3 Effect of Charging Number of Fine Particles

In addition to the friction coefficient, the concentration of fine particles is also a crucial parameter in determining whether migrating fine particles will clog. In general, clogging occurs more easily at higher concentrations of fines. Fig. 14-6 shows the distribution of clogging particles under different charging number. One can observe that the deposited fine particles tend to develop far away from the center with the increase of the number of charging fines. As can be seen in Fig. 14-6(a), few fine particles are deposited in the packed bed, and there is no obvious cluster blockage. These discrete clogging particles are mainly due to friction between one fine particle rolling very slowly on the surface of large particles whose spacing is close to the diameter of small grains. With the increase of charging number of fine particles, the more fine particles are deposited in the packed bed, especially the number and size of cluster increases obviously. In Fig. 14-6(d), with the charging number of fine particles $N=1500$, the main of the blockage is due to the collective interactions between percolating particles and is responsible for the clogging of pores in the granular system.

(a)　　　　　　　(b)　　　　　　　(c)　　　　　　　(d)

Fig. 14-6　Influence of charging number of fine particles on the distribution of blockage

(a) $N=100$; (b) $N=500$; (c) $N=1000$; (d) $N=1500$

Fig. 14-7 shows the variation of the clogging ratio with the charging number of fine particles. When the fine particles settled in the packed bed reaching stability, with the charging of fine particles increase from 100 to 1500, the number of deposited particles increase from 14 to 775, and the corresponding clogging ratio increases from 14% to 51.7%. The main reason is that with an increase of the charging number, the mass flow rate in an orifice increases, which indicating the frequency of simultaneous falling of multiple fine particles into the orifice increases. During passing through the channel, the fine particles experience more collisions with the each other, and the reduction of powder momenta caused by collision increases. As shown in Fig. 14-8, at stable stage, the contact number between the fine and fine particles increases from 28 to 990. Consequently, in a fix time interval, the existing number of fine particles on the orifice increases with the increase of mass flow rate, which will generate more chance of fine particles creating arch between packed particles. In the actual process of pulverized coal injection in blast furnace, the amount of unburned pulverized coal should be reduced by strengthening pulverized coal combustion, thereby decreasing the mass flux of fine particles at the borders of raceway and reducing the fine blockage. For reducing the frequency of the chocking of slot in COREX shaft furnace, one effort should be made to improve the efficiency of hot gas cyclone to reduce the fines content in the reducing gas entering the shaft furnace.

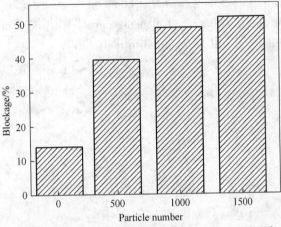

Fig. 14-7 Influence of charging number of fine particles on the clogging ratio

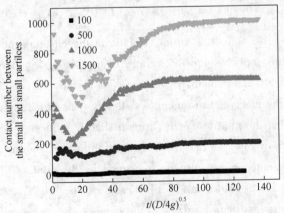

Fig. 14-8 Influence of charging number of fines on the contact number between the percolation fine particles

14.1.4 Effect of Initial Velocity

In COREX shaft furnace, due to the install of AGD in the bustle zone, the inlet gas velocities of reducing gas at different slots are different, and thus the initial velocities of fine particles carried by reducing gas are different. This work discussed the influence of initial velocity of fine particles on the distribution of blockage, as shown in Fig. 14-9. The velocity is in units of free fall velocity reached after falling over one packed particle. It can be seen that the distribution characteristics of the clogging fine particles are basically the same, except that the size of the cluster at the top of the packed bed reduces slightly with the increase of the initial velocity. This means that under the condition of the same bed structure, the initial velocity has minor effect on the distribution of the clogging fine particles. Impact on the distribution of clogging particles in the context of larger initial velocity is in progress and will be reported in the future.

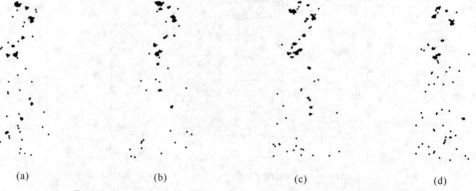

(a)   (b)   (c)   (d)

Fig. 14-9 Influence of initial velocity on the distribution of blockage

(a) $v=0$; (b) $v=(2gD)^{0.5}$; (c) $v=(4gD)^{0.5}$; (d) $v=(6gD)^{0.5}$

Fig. 14-10 shows the effect of initial velocity on the clogging ratio. When the initial velocity of the fine particles increases from 0 to $(6gD)^{0.5}$, the clogging ratio is 39.4%, 39.6%, 40.2% and 37% respectively. In the range of initial velocity less than $(4gD)^{0.5}$, the clogging ratio increases slightly. This is mainly due to the increase of the initial vertical downward velocity, which slightly reduces the radial diffusion behavior of the fine particles, so that the frequency of dust collision in the orifices slightly increases. The phenomenon can also be explained by the contact number between the fine and fine particles as shown in Fig. 14-11. As can be seen from Fig. 14-11, when the initial velocity is less than $(4gD)^{0.5}$, the number of particle collisions at the speed of $(4gD)^{0.5}$ at different times is the largest, reflecting the increase in the deposition rate of fine particles. When the initial velocity reaches $(6gD)^{0.5}$, although some bottleneck particles form arches on packed bed and stop the flow, the kinetic energy of the following fine particles are large, which will destroy the bridging effect, thus reducing the dust deposition rate. Thus, in the design of bustle pipe of COREX shaft furnace, the reduction of the cross-sectional area of the bustle pipe and slot is an effective way to improve the gas-fine velocity and thereby reduce fine particles blockage in packed bed in front of the slot. It can also be seen from Fig. 14-9(d) that some cluster structures disappear compared with other initial velocity conditions.

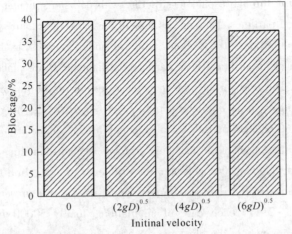

Fig. 14-10 Influence of initial velocity on the clogging ratio

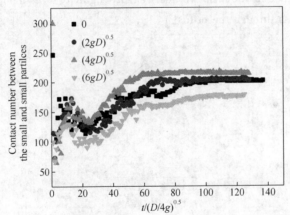

Fig. 14-11 Influence of initial velocity on the contact number between the percolation fine particles

14.2 Influence of Cohesive Force

14.2.1 Simulation Conditions

In the actual packed bed of COREX shaft furnace, the burden materials having irregular shapes are randomly packed. It is considered that the narrow part of the flow path is important to discuss the powder flow and accumulation in the packed bed. In this work, a simplified packing structure, namely an orifice consisting of three spherical particles that touch each other and are arranged in an equilateral triangle, is used for simplification. In the practice, the fine particles are entrained by the reducing gas and flows mainly upward. This study focuses on the fundamental blockage behaviour of cohesive fine particles moving through a packed structure with collisions among the fine particles and with the packed particles. Thus no gas flow in the packed structure is considered and the small fine particles pass the packed structure under gravity. Fig. 14-12 shows the schematic diagram of the orifice. These particles of diameter D are placed horizontally at the height 0.01m. The fine particles of diameter d are put on the top of the orifice. They are generated randomly at an equilateral triangle of length D. The height of the triangle above the centre of packed particle is also equal to the diameter of packed particle D. The dropping interval of fine particle is given at constant time interval of $t = 0.02$s. The fine particles are dropped onto the orifice by gravity, and the trajectories of the particles are numerically tracked. The parameters used in the present simulations are listed in Table 14-2. For each case of simulation, totally 50000 fine particles are dropped and the frequency of the blockage par 50000 particles is recorded. The descending velocity is defined as the mean vertical velocity (gravity direction) of the fine particles which can pass through the orifice.

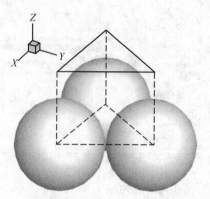

Fig. 14-12 Geometry of the orifice used in this work

Table 14-2 **Particle properties and simulation conditions**

Variables	Value	Variables	Value
Diameter of packed particle D/m	0.01	Poisson's ratio v_p	0.3
Powder diameter d	0.105~0.13D	Damping coefficient c	0.3
Density ρ/kg·m^{-3}	2000	Time step Δt/s	1.0×10^{-6}
Sliding frictional coefficient μ_s	0.3	Mass flow rate Q ($g\pi d^3 \rho/6/T$)	1~3
Young's modulus E/Pa	1×10^7	External forces f_g ($g\pi d^3 \rho/6$)	0~0.6

14.2.2 Blockage Formation and Mechanism

Fig. 14-13 shows the evolution of height and velocity of one representative fine particle under different sticking forces conditions when $d/D=0.13$ and $f_g=0$. In this figure upward velocity is shown as positive value. It is indicated that at the very beginning, the height evolution shows a parabolic curve and the velocity increases rapidly due to the free fall. Once the fine particle collides to the packed large particle, it receives upward contact force from the packed particle and the velocity decreases rapidly, as shown in Fig. 14-13(b). Then the fine particle would move toward the opening of the orifice along the wall and enters the space between two packed particles. For those fine particles

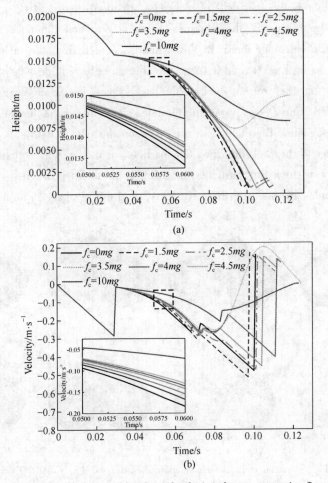

Fig. 14-13 Evolution of height and velocity of a representative fine particle for different sticking force f_c when $d/D=0.13$ and $f_g=0$
(a) Height; (b) Velocity

with different interparticle stickiness, after experiencing several collisions, those fine particles with smaller interparticle stickiness ($0 \sim 4mg$) can pass the narrowest part (pore throat) of the orifice and drops to the bottom, while those fine particles with higher stickiness ($\geqslant 4.5mg$) will finally stick on the surface of the packed bed. With the increase of the sticking force between powders and packed particles, the fine particles are more difficult to move toward the bottom, and the descending height and velocity of fine particles decrease. Especially, when the stickiness is increased over a certain level ($\geqslant 4.5mg$ in this case), the sticking force is strong enough to resist the inertia motion of fine particles and the fine particles will adhere to the surface of the packed particles.

Fig. 14-14 shows the snapshots of blockage formation when $d/D=0.13$, $f_c=0$, $Q=1$ and $f_g=0$. In this case, the stickiness force is zero and the interval between the figures is 0.02s. The time shown in the figure is counted from the moment when four particles exit the orifice. At time 0.02s, two fine particles collide near the narrow part of the orifice, but they are not stuck on the packed particles. So no clogging occurs during $0 \sim 0.06$s. At time 0.08s, two fine particles simultaneously enter the narrow part of the orifice at the same time and both are stuck on the packed particles. The third fine particle collides to the preceding two fine particles at 0.10s. This collision brings little motion of the two stuck fine particles, and finally these fine particles settle there. The following particles are piled on these particles, and the blockage is formed.

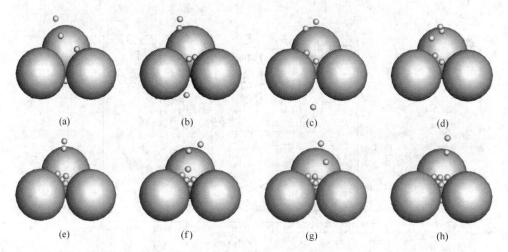

Fig. 14-14　Snapshots of blockage formation when $d/D=0.13$, $f_c=0$, $Q=1$ and $f_g=0$
(a) $t=0$s; (b) $t=0.02$s; (c) $t=0.04$s; (d) $t=0.06$s;
(e) $t=0.08$s; (f) $t=0.10$s; (g) $t=0.12$s; (h) $t=0.14$s

As observed in Fig. 14-13, the fine particles with higher stickiness (e. g. $10mg$) will finally stick on the surface of the packed bed. The formation process of the orifice blockage by fine particles for the sticking force of $10mg$ is shown in Fig. 14-15. The other parameters are the same as those in Fig. 14-14. At time 0s, five fine particles are used in the calculated domain including three located above the narrowest part of the orifice and the other two adhered to the surface below to the narrowest part of the packed particle. At time 0.06s, four more fine particles pass the pore throat and stick on the surface of the packed particles. All these trapping fine particles can effectively reduce the sectional area of the flow channel, and the following fine particles directly settle above these trapped fine particles. Then the clogging can be observed after 0.08s.

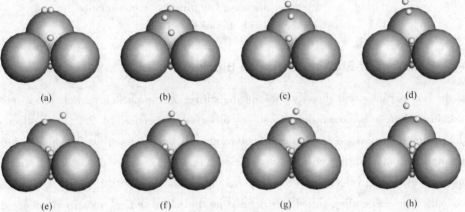

Fig. 14-15 Snapshots of blockage formation when $d/D=0.13$, $f_c=10mg$, $Q=1$ and $f_g=0$
(a) $t=0s$; (b) $t=0.02s$; (c) $t=0.04s$; (d) $t=0.06s$; (e) $t=0.08s$;
(f) $t=0.10s$; (g) $t=0.12s$; (h) $t=0.14s$

Based on the above simulation results, two kinds of blockage mechanisms are identified. In the case without interparticle stickiness, the mechanical is the same as described in section 14.1.2. In the case with high sticking force between fine particles and packed particles ($10mg$), the fine particles passage behaviour is terminated and they can finally stick on the surface of the packed bed. The fine particles adhere to the surface near the pore throat of the orifice acting as bottleneck of void space and could directly decrease the sectional area of the flow channel. When the equivalent diameter of the flow channel is decreased to a critical level that the following fine particles could not pass through, the blockage is formed. The fine particles deposition in packed bed with higher sticking force is schematically shown in Fig. 14-16. It should be pointed out that, mechanism that powders bridge across the pore throat also contribute to the clogging of fine particles with higher interparticle stickiness.

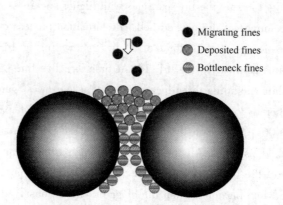

Fig. 14-16 Schematic diagram of fine particles deposition in the packed bed with sticking force $10mg$

14.2.3 Effect of Sticking Force on Blockage

As the fine particles will directly stick on the surface of the packed particle when interparticle stickiness is larger than $4mg$, as observed in section 14.2.2, the effect of sticking force on passage and blockage behaviour will be discussed for the sticking force ranging from $0mg$ to $3mg$ only in this section. Fig. 14-17 shows the effect of sticking force on descending velocity and blockage frequency when $d/D=0.13$, $Q=1$, $f_g=0$. In this part, the descending velocity is defined as the mean vertical velocity (gravity direction) of fine particles during passing through the orifice (from height 0.015m to 0.005m). The particles that cannot pass the orifice due to the blockage are eliminated. The fine particles are dropped at the height 0.02m and the residence time and descending velocity of free falling from height 0.015m to 0.005m are 0.0232s and 0.432m/s, respectively. The descending velocity of the fine particles without interparticle stickiness is just 0.223m/s. As the area ratio of the pore throat of the orifice to the powder generating area is about 9.31%, most fine particles collide on the packed particles and descending along the surface. Consequently, the descending velocity shows almost half of the free fall descending velocity. With the increase of sticking force, the descending velocity decreases. The sticking force can resist the inertia motion of fine particles and hence increase the residence time. Longer residence time causes more possibility for the collision with other fine particles. Hence, the larger the sticking force is, the smaller will be the descending velocity. The effect of sticking force on the frequency of blockage par 50000 particles is also shown in Fig. 14-17. The increase of sticking force weakens the bounce of fine particle after collision. The attenuation of particle kinetic energy becomes stronger, and the frequency of simultaneous falling of multiple fine particles into the orifice increases. As a result, the possibility of orifice blockage increases with the increase of the sticking force.

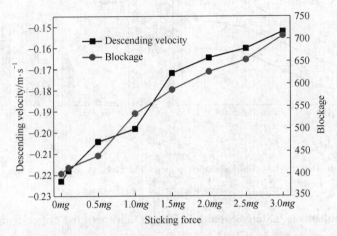

Fig. 14-17 Effect of sticking force on descending velocity and blockage frequency when $d/D=0.13$, $Q=1$ and $f_g=0$

14.2.4 Effect of Other Key Variables on Passage and Blockage Behaviour

Fig. 14-18 shows the effect of size ratio of fine particle to packing particle on passage behaviour. It is indicated that with increasing the particle size ratio, the descending velocity decreases, as shown in Fig. 14-18(a). This can be easily explained by considering a single particle falling down toward the orifice. The particle trajectory must be aligned with the pore hole to pass through it without bouncing around for small ratio of particle size. If not, when size ratio is larger, the collisions of particles will be more significant. As such, the downward motion of the fine particle is more difficult and leads to a decrease in the descending velocity. Fig. 14-18(b) shows the effect of particle size ratio on the frequency of blockage. The frequency of blockage increases with the increase of the particle size ratio. With increase of fine particle diameter, the ratio of projection area of fine particle to the area of the orifice opening increases, and the blockage of the orifice opening by a few fine particles bridging across will be more possible.

Fig. 14-19 shows the effect of mass flow rate on passage behaviour of powder. In this part, the dropping interval of fine particle is given at constant time interval $t=0.02s$, and the mass flow rate Q represents the number of fine particle generated at each interval. From Fig. 14-19(a), one can see that the higher the mass flow rate is, the smaller are the descending velocities. With an increase of the mass flow rate, the fre-

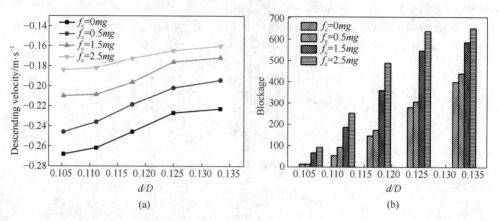

Fig. 14-18　Effect of size ratio on descending velocity and blockage frequency when $Q=1$ and $f_g=0$

(a) Descending velocity; (b) Blockage

quency of simultaneous falling of multiple fine particles into the orifice increases. During passing through the orifice, the fine particles experience more collisions with the each other and the reduction of powder momenta caused by collision increases. Consequently, the descending velocity decreases with the increase of mass flow rate. At the same time, from Fig. 14-19(b), it can be seen that the frequency of blockage increases with the increase of mass flow rate. Then in a fix time interval, the existing number of fine particles on the orifice increases with the increase of mass flow rate, which will generate more chance of fine particles creating arch between packed particles.

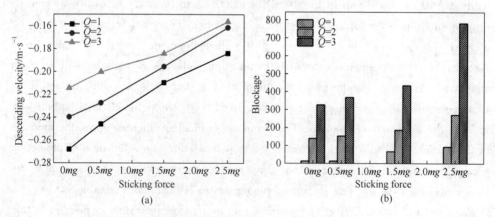

Fig. 14-19　Effect of mass flow rate on descending velocity and
blockage frequency when $d/D=0.105$ and $f_g=0$

(a) Descending velocity; (b) Blockage

Counter-current gas flow is a typical practice in a packed bed of shaft furnace in

COREX technology. In this case, the particles are subjected to an external force. Such a force (upward direction) will contribute to the movement of particles, and hence influence the blockage behaviour. In practice, the gas drag force on each particle depends on the gas flow distribution corresponding to the local permeability. However, for brevity, it is assumed that the particles are only affected by a constant force in upward direction, f_g, in addition to contact forces, gravity and cohesive force.

Fig. 14-20 shows the effect of external force on the passage behaviour of fine particles. From Fig. 14-20(a), it can be seen that, with increasing the external force, the descending velocity decreases. The main reason is due to the vertical acceleration of the fine particles. For high vertical force applied in the opposite direction of the gravity, the acceleration is smaller and the residence time is longer. Especially, when the $f_g=0.6mg$ and $f_c=2.5mg$, the fine particle cannot passage through the orifice. As shown in Fig. 14-13, when the ratio of stickiness to gravity is increased over 4.5 in this work, the fine particle can adhere to the surface of the packed particle. For the case with $f_g=0.6mg$ and $f_c=2.5mg$, the ratio of stickiness to the net gravity is 6.25, which is larger than the critical level. So the descending velocity for this condition is not considered in Fig. 14-20(a). The effect of the external force on the frequency of blockage is shown in Fig. 14-20(b). It can be observed that with the increase of external force, the frequency of blockage increases. As discussed above, the higher the external force is, the longer will be the residence time. So the existing number of the fine particles on the orifice will increase, which is benefit for the bridging action of fine particle. Besides, the high vertical force could lead to a larger ratio of stickiness to the net gravity, which causes the attenuation of particle kinetic energy becoming stronger.

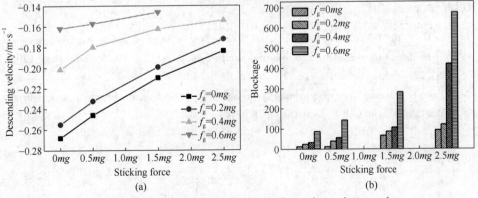

Fig. 14-20 Effect of external force on descending velocity and blockage frequency when $d/D=0.105$ and $Q=1$
(a) Descending velocity; (b) Blockage

14.3 Summary

The blockage behavior of fine particles in a packed bed of large sphere particles is studied numerically by means of DEM approach. The following conclusions are obtained:

Fine particles blockage in packed bed is easy to occur in the area where fine particles just enter the bed. The fine particles with larger rolling friction coefficient have a higher clogging ratio. The charging number of fine particles directly affects the blockage. With the increase of charging fine particles, the fine particles experience more collisions with the each other, and the clogging ratio increase. When the initial velocity is between 0 to $(4gD)^{0.5}$, the deposition rate increases slightly with the increase of initial velocity, while when the velocity increases to $(6gD)^{0.5}$, the clogging ratio decreases.

With the increase of sticking force, the descending velocity will decrease, while the frequency of blockage increases. There are two mechanisms of blockage formation. One is without interparticle stickiness. It is mainly due to mechanical interactions between fine particles, which can create arches on packed bed and stop the flow. The other is that, when interparticle stickiness is very high (e. g. $10mg$), fine particles can finally stick on the surface of the packing particles. The fine particles adhere to the surface near the pore throat of the orifice acting as bottleneck of void space and can directly decrease the sectional area of the flow channel. When the equivalent diameter of the flow channel is decreased to a critical level that the following fine particles can not pass through, the blockage is formed. The size ratio of fine particle to packing particle, mass flow rate and external forces can directly affect the blockage behaviour of fine particle. Increasing the size ratio, mass flow rate and external force will increase the frequency of blockage.

References

[1] Fan H J, Mei D F, Zhang F G, et al. DEM simulation of different particle ejection mechanisms in a fluidized bed with and without cohesive interparticle forces [J]. Powder Technology, 2016, 288: 228~240.

[2] Civan F. Reservoir formation damage: fundamentals, modeling, assessment, and mitigation [M]. 2nd edition. Gulf Professional Pub. Elsevier, 2007.

15 CFD-DEM Study of Fine Particles Behaviors in a Packed Bed with Lateral Injection

The transport of fines in suspension flow through a porous media may happen in many industrial processes, of particular interests to ironmaking process. For example, COREX, a smelting reduction process, is a new, cost-efficient, and environmental-friendly technology, as less coke is used. Two-stage procedures are included in COREX involving pre-reduction in a shaft furnace, followed by final reduction and separation in a melter gasifier. The powders or fine particles generated from the melter gasifier are entrained by the reducing gas stream. The large fines are removed in a hot gas cyclone, then the reducing gas with a dust load of $20g/m^3$ is blasted into the shaft furnace at low velocity (8~20m/s) through the slot in bustle pipe zone. The distribution and clogging behaviors of fines can impact the voidage of the packed bed and narrow the flow paths among the packed particles. This directly affects the smoothness and performance of the shaft furnace. Thus, it is important to analyze the fine particles behaviors in a packed bed, with the target to prevent or reduce the impact of this phenomenon.

In this work, the fine particles clogging behavior in a packed bed with lateral inlet is investigated by means of CFD-DEM. The flow and clogging of powders of varying critical variables were studied. The clogging mechanisms were also discussed. The findings of this work provides a fundamental understanding on clogging behavior of powders in a packed bed with lateral inlet.

15.1 Simulation Conditions

The three-dimensional calculation region is shown in Fig. 15-1. The simulation domain which is a rectangular container of $10D \times 10D \times 20D$ filled with a packing of monosize large sphere particles of diameter D. The geometry is divided into hexahedral mesh, and the length of the square grid equal to $2D$. The porosity of the packed bed is around 0.4. A fixed amount of powders is mixed with the gas that flows from the lateral inlet. The inlet is located at the center of the left side and 0.04m from the bottom. The detailed numerical conditions are listed in Table 15-1. The update interval of the gas flow distribution was set to once every 10 steps of the DEM model, based on the convergence property and the accuracy of the calculation. Alumina and coal were used as packed particles and fines, respectively. The total simulation time is 2.0s. In the first 1s, 5000 powder particles are injected into the packed bed with airflow, and only gas was injected in the last 1s.

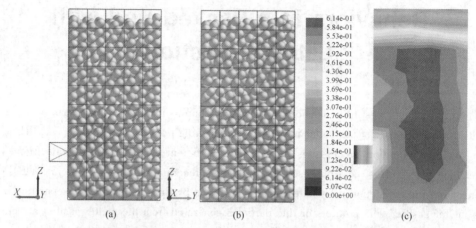

Fig. 15-1 Schematic diagram of the calculation domain in this work
(a) Section in y direction; (b) Section in x direction; (c) Void distribution

Table 15-1 Parameters used in the simulation

Particle	Packing particle	Fine particle
Diameter/mm	10	0.5, 0.625, 0.75, 0.875, 1(Base case)
density/kg · m^{-3}	3.95×10^3	1.10×10^3
Shear modulus E/Pa	2.1×10^{10}	1×10^7
Poisson's ratio v_p	0.24	0.21
Restitution coefficient e	0.2	0.2
Sliding frictional coefficient μ_s	0.5	0.5
Rolling frictional coefficient μ_r	0.01, 0.02(Base case), 0.04, 0.06, 0.08	
Mass flux f/kg · m^{-2} · s^{-1}	1.44, 4.32, 7.20(Base case), 10.08	
Velocity/m · s^{-1}	20(Base case), 40, 60, 80	
Time step/s	1.0×10^{-6}	
Fluid's density/kg · m^{-3}	1.2	
Fluid's viscosity/Pa · s	1.8×10^{-5}	
Fluid's velocity/m · s^{-1}	20	
Fluid's time step/s	1.0×10^{-5}	

15.2 Model Validity

The CFD-DEM method is validated by means of comparison of the pressure drop in a packed bed within spherical particles. Fig. 15-2 shows the comparison of the pressure drop obtained from the Ergun correlation and CFD-DEM simulation under different particle diameters. The height of the packed bed under different particle diameters keeps the same, and the average voidage of the packed beds is around 0.386. The gas flow with 1m/s is introduced from the bottom distributor. As the particle diameter increases from 0.06m to 0.14m, both the simulation and calculation results decrease, and the simulated and theoretical pressure drops decrease from 1029Pa and 1011.8Pa to 451Pa and 430.4Pa, respectively. The simulated pressure drops are basically consistent with the theoretical calculation results, which confirms the feasibility and rationality of the current CFD-DEM method in predicting gas-solid flow behaviours in a packed bed with lateral injection.

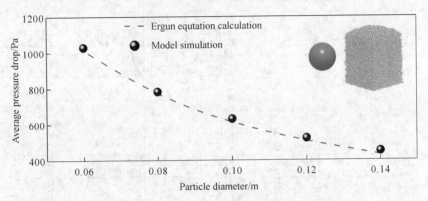

Fig. 15-2 Comparison of pressure drop between simulation and theoretical calculation

15.3 Results and Discussion

15.3.1 Effect of Gas Velocity

Snapshots of the powder distribution at different time intervals are shown in Fig. 15-3. When $t=0.01s$, the powders just enter the porous media, and the penetration depth of the dust in the horizontal direction increases with the increase of the blowing speed. With the continuous injection of powders, it can be found that there are dynamic powders at the side wall near the entrance. The main reason is due to the change of the packed bed caused by the restrained packed particles with wall, and the packed bed structure near the wall has good permeability. In addition, a large amount of powder deposited at the bottom of the packed bed can also be found. This is mainly because part of powders collide with the packed particles, which changes the force behavior, and the downward force is greater than the upward drag force. Influence of the velocity on the powders deposited at the bottom will be discussed in detail later. When $t=1s$, the maximum hold up of powders in the voids of packed bed can be observed at different gas velocity. Since the addition of powder stopped after 1s, the hold up of powders in packed bed decreased and reached a stable state at 2s. There are two kinds of clogging powders inside the porous. One is a cluster composed of agglomerated powders, the other is a dispersed single powder. Most of the clogging powders are located at the same level of the entrance. In the final stable state, with the increase of the gas velocity, the clusters decreases, while the dispersed powders increases, especially at the height above the powder inlet.

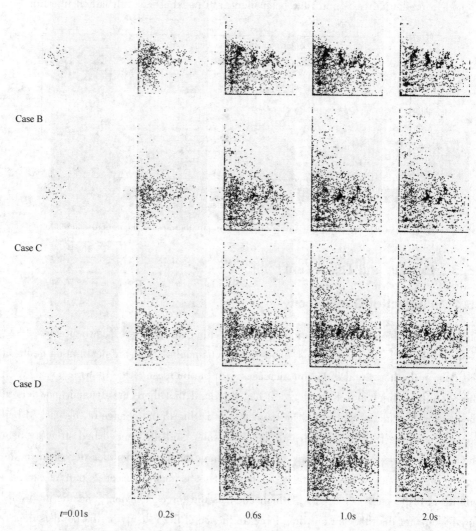

Fig. 15-3 Snapshots of powders distribution in packed bed under different velocity, Case A 20m/s, Case B 40m/s, Case C 60m/s, Case D 80m/s

Fig. 15-4 shows the influence of gas velocity on the ratio of powders settling at the bottom of the packed bed. When the gas velocity increases from 20m/s to 80m/s, the proportion of powders settling at the bottom of packed bed decreases from 43.62% to 29.96%. The main reason is expected that the higher the gas velocity, the larger the upward drag force on the powder particles. Thus, the probability of powders leaving the gas streamline after collision with packed particle and settling at the bottom lower decrease under high gas velocity. The following discussion of the distribution of the hold up of powders in packed bed will not include these powders settling at the bottom.

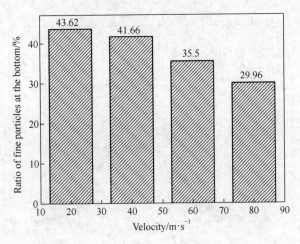

Fig. 15-4   Effect of velocity on the proportion of powders settling at the bottom

In this work, the mass of powders in every 0.002m interval in y or z direction is counted to characterize the distribution of hold up. The transverse (y direction) distribution of hold up of powders in packed bed is shown in Fig. 15-5(a). It can be found that the transverse distribution at different velocities has the characteristics similar to normal distribution. With the increase of gas velocity, the more powders diffuse in the transverse direction. In practical production, such as COREX shaft furnace, the main reason for the chock of gas slot is the initial powders deposition zone formed in front of the inlet. The main area of powder clogging is around the gas-powder inlet. Therefore, we further discuss the distribution of hold up of powders in the range of the inlet size. As shown in the upper right corner of Fig. 15-5(a), when the gas velocity increases from 20m/s to 60m/s, the powders are easy to move to the wall, so that the number of powders at the inlet level decreases with the increase of gas velocity. However, when the velocity increases to 80m/s, the proportion of powders settling at the bottom of packed bed is lower, which makes the amount of powders deposited at the inlet level increase. From the proportion of deposited powders at inlet level, it can be found that the proportion decreases with the increase of gas velocity. Fig. 15-5(b) shows the longitudinal (z direction) distribution of hold up of powders in packed bed. It can be found that the amount of deposited powder in the packed bed decreases gradually with the increase of gas velocity in the region below 40mm, while it is opposite in the area higher than 60mm. This phenomenon is consistent with that in Fig. 15-3. At the inlet level zone in longitudinal direction, both the number and proportion of deposited powders decrease with the increase the gas velocity.

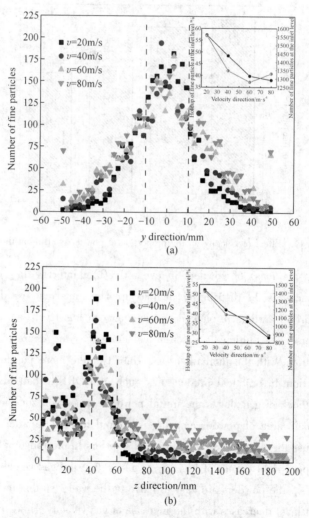

Fig. 15-5 Distribution of hold up of powders in the packed bed under different velocity
(a) y direction (transverse direction); (b) z direction (longitudinal direction)

The streamline of the gas flow is shown in Fig. 15-6. The distribution of the streamline is obtained in the steady state of the gas flow in packed bed ($t=2$s). The gas distribution injected from the lateral inlet of the packed bed shows a 'J' shape for all cases. It is expected that the gas velocity in packed bed increases with the increase of injecting speed. Combined with the powder distribution in Fig. 15-3, it can be seen that most of the powders are separated from the gas streamlines. The smaller the gas velocity, the lower the degree of agreement between the powder movement and the gas streamline.

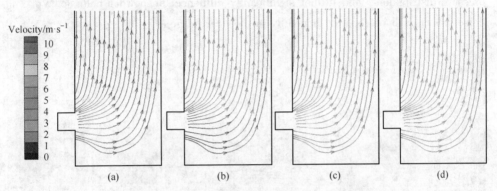

Fig. 15-6 The streamline of the gas flow in packed bed
(a) 20m/s; (b) 40m/s; (c) 60m/s; (d) 80m/s

Fig. 15-7 shows the evolution of pressure drop with time under different velocity. With the continuous injection of powders, the pressure drop gradually increased and reached the maximum value in 1s. When the injecting velocity increases from 20m/s to 80m/s, the maximum pressure drop increases from 6935.7Pa to 34541.8Pa. Within 1~2s, no

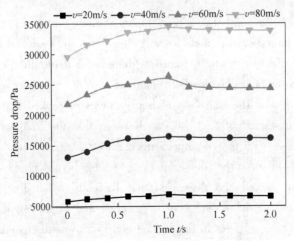

Fig. 15-7 Evolution of pressure drop with time under different velocity

powder was injected and the pressure drop of the packed bed decreases and tends to be stable. Fig. 15-8 shows the effect of injection velocity on the constant pressure surface when $t=2s$. The absolute value of pressure in Fig. 15-8 is shown as a curved surface of the same color, and the pressure drop increases as the distance between the isobar planes reduces. The amount of pressure increases as the injection velocity increases. The change in the horizontal direction is also can be seen under different gas velocity. General speaking, in the actual BF process, the cohesive zone could affect the gas and pressure distribution, and further significantly influence the powder hold up. Work on these aspects will be reported hopefully in the near future.

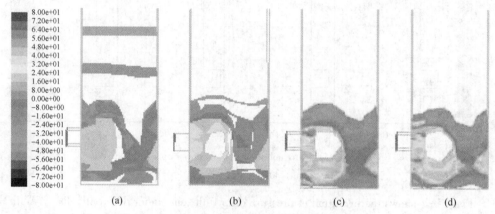

Fig. 15-8 Constant pressure surface in packed bed when $t=2s$
(a) 20m/s; (b) 40m/s; (c) 60m/s; (d) 80m/s

15.3.2 Effect of Diameter Ratio

The geometrical relationship between pore throats and fine particles is crucial in determining whether the migration fine particles will clog or not. In a packed bed, the geometrical ratio of pore throat width to fine particle diameter is determined by the size ratio of fine particle diameter to packed particle diameter (d/D). In this work, the variable varies from 0.05 to 0.1. The snapshots of fine particles in packed bed under $d/D=0.1$ can be observed in Fig. 15-9(a). It can be seen that the fine particles clogging in packed bed increases with the continuous injection of powders. When $t=1s$, the maximum hold up of powders in the voids of packed bed can be observed. After 1s, the hold up of powders in packed bed decreased and basically reached a stable state after 1.4s. Thus, the fine particles were actually trapped and would not transport further at 2s. The hold up of fine particles in the packed bed $t=2s$ was discussed. Fig. 15-9(b) ~ (e) shows the distribution of the fine particles at $t=2s$ under different diameter ratios. For the case of $d/D=0.1$, some fine particles concentrate at the sidewall near the entrance. The main reason is the change of the packed bed caused by the restrained packed particles with the wall, and the structure of the packed bed near the wall has good permeability. Besides, a large amount of fine particles deposit at the bottom of the packed bed. This is mainly due to part of fine particles collide with the coarse packed particles, which leads to the downward force greater than the upward one. There are two kinds of clogging powders inside the porous media when the $d/D=0.1$. One is a cluster composed of agglomerated powders, and the other is a dispersed single powder. Most of the clogging fine particles are located at the same latitude as the entrance. With the decrease of diameter ratio, the number of fine particles at the sidewall near the entrance decreases. With the decrease of fine particle diameter, the cluster in the packed bed decreases while the dispersed fine particles increase. The smaller fine particle is easily carried to the upper region of the packed bed by the gas flow and even blown out of the packed bed.

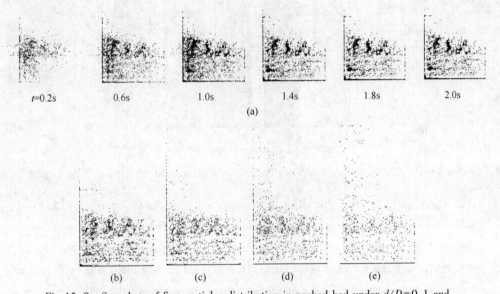

Fig. 15-9 Snapshots of fine particles distribution in packed bed under $d/D=0.1$ and distribution of fine particles at $t=2s$ under different diameter ratios
(a) Snapshots of fine particles distribution in packed bed under $d/D=0.1$; (b) $t=2s$, $d/D=0.0875$; (c) $t=2s$, $d/D=0.075$; (d) $t=2s$, $d/D=0.0625$; (e) $t=2s$, $d/D=0.05$

Fig. 15-10 shows the effect of diameter ratio on the proportion of fine particles settling at the bottom of the packed bed. With the increase of diameter ratio, the fraction of fine particles at the bottom increases first and then decreases. When the diameter ratio increases from 0.05 to 0.075, the proportion of fine particles settling at the bottom of the packed bed increases from 46.6% to 52.11%. This is mainly because the smaller diameter ratio results in smaller gravity and larger upward force. Thus, the probability of fine particles leaving the gas streamline after collisions with packed particles and settling at the bottom decreases under the smaller diameter ratio. However, when the diameter ratio increases from 0.075 to 0.1, the proportion of fine particles at the bottom decreases to 43.61%. The main reason is that the dispersion of large fine particles carried by the inlet gas in the packed bed is poorer than that of small fine particles, and the large ones are more likely to deposit in the voidage of the packed bed rather than at the bottom. The following discussion of the distribution of the hold up of fine particles in the packed bed will not include this settling at the bottom.

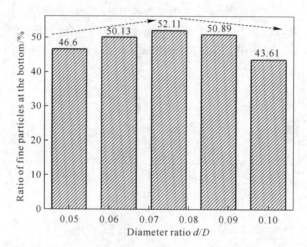

Fig. 15-10　Effect of diameter ratio on the proportion of fine particles settling at the bottom

Fig. 15-11(a) presents the effect of diameter ratio on the transverse (y direction) distribution of hold up of fine particles in the packed bed. The transverse distributions at different diameter ratios show similar patterns to the normal distributions. In practical production, with the increase of PCI rate, the unburnt char and the fine coke may accumulate in the boundary of the raceway and particularly beyond the end of raceway at the tuyere latitude, forming a low-permeability region, termed as bird's nest. The chock of the gas slot in the COREX shaft furnace is mainly due to the fine particles clogging around the gas-fine inlet. Therefore, we further discuss the distribution of hold up of fine particles in the range of the inlet size ($-10\text{mm}<y<10\text{mm}$ or $40\text{mm}<z<60\text{mm}$). As shown in the upper right corner of Fig. 15-11(a), the number of static fine particles at the inlet latitude is the largest with $d/D=0.1$, while it is the smallest with $d/D=0.075$. Usually, larger diameter ratios cause the pore throats acting like gates in the packed bed, which will create a 'gate or valve effect' and lead to an impedance on the transverse motion of fine particles. Therefore, when the d/D increases from 0.075 to 0.1, the number of static fine particles at the inlet latitude increases. However, affected by the drag force of the gas, when the d/D decreases from 0.075 to 0.05, the fine particles carried by the gas flow mainly move upwards, and the transverse motion becomes weaker, so the static fine particles at the inlet level increases. The longitudinal (z direction) distribution of hold up of fine particles in the packed bed is shown in Fig. 15-11(b). Along with the z direction, the maximum number of static fine particles under different diameter ratios appears at the inlet latitude (40mm< z <60mm). Both the number and proportion of static fines increase with the increase of the diameter ratio.

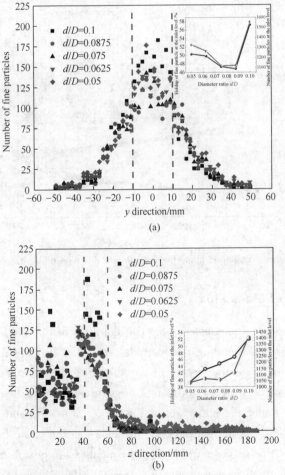

Fig. 15-11 Effect of diameter ratio on the distribution of hold up of fine particles in the orifice of the packed bed

(a) y direction (transverse direction); (b) z direction (longitudinal direction)

Fig. 15-12 shows the effect of diameter ratio on the pressure distribution in the packed bed along with the longitudinal direction. The data was taken from the position where $y = 0$mm and $x = 90$mm (10mm away from the gas injection). The maximum pressure appears at $z = 60$mm for all cases. The maximum pressure increases from 769.7Pa to 949.5Pa when the diameter ratio increases from 0.05 to 0.1. It is expected that large fine particles deposited in front of the inlet are in a higher number and occupy more void volume, which increases the pressure drop of gas. In the area higher than 60mm, the diameter ratio has a minor effect on the pressure distribution due to the less fine particles deposited and a more intensive dispersion.

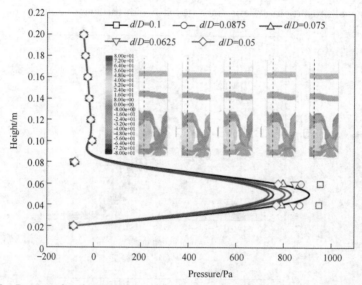

Fig. 15-12 Pressure distribution along the longitudinal direction under different diameter ratios

15.3.3 Effect of Mass Flux

In addition to geometrical parameters, the concentration of fine particles is also crucial in determining whether migrating fine particles will clog. In general, clogging occurs more easily at higher concentrations of fine particles. Fig. 15-13 shows the distribution of fine particles at a stable state under different mass fluxes. Most of the clogging fine particles are dispersed under the mass flux of 1.44 kg/(m² · s), and a small cluster is formed near the middle of the packed bed, as shown in the circle in Fig. 15-13(a). As the mass flux of fine particles increases, the number and probability of fine particles entering the same orifice increase, which is easier to form agglomerated clogging fine particles. The main reason is that during fine particles passing through the channel, with the increase of frequency of simultaneous falling of multiple fine particles into the orifice, the fine particles experience more collisions with the each other, and the reduction of powder momenta caused by collision increases. As shown in Fig. 15-13(d), the number of clusters increases significantly, and the size of clusters further increases. Compared with the cluster in the circle in Fig. 15-13(a), the size is obviously developed, and its shape is elongated with the increase of mass flux.

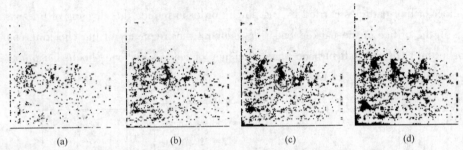

Fig. 15-13 Distribution of fine particles under different mass fluxes
(a) 1.44kg/(m² · s); (b) 4.32kg/(m² · s);
(c) 7.20kg/(m² · s); (d) 10.08kg/(m² · s)

The influence of mass flux on the proportion of fine particles settling at the bottom is shown in Fig. 15-14. The mass flux of fine particles increases from 1.44 to 10.08kg/(m² · s), and the proportion of fine particles settling at the bottom increases from 38.9% to 44.75%. The collision frequency between fine particles and coarse packed particles increases as the mass flux of fine particles increases. The fine particles in the packed bed with high mass flux may lose kinetic energy after colliding with the coarse packed particles. Consequently, the velocity of these fine particles may be less than the critical transport/conveying velocity for fine particles, which makes it easier to deposit at the bottom of the bed.

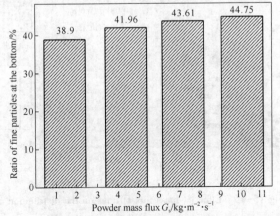

Fig. 15-14 Effect of mass flux of fine particles on the proportion of
fine particles settling at the bottom

Fig. 15-15 presents the distribution of hold up of fines in the packed bed under different mass fluxes. Both the transverse and longitudinal directions show that with the increase of mass flux, the number of static fine particles in the orifice of the packed bed gradually increases. The inverted triangle with the maximum mass flux are at the top of all conditions. However, after normalization, the proportion of static fine particles decreases with the increase of mass flux at the inlet latitude. It is necessary to reduce the number of fine particles carried by gas flow in order to reduce the clogging of fine particles in the orifice of the packed bed. For reducing the frequency of the chocking of the slot in the COREX shaft furnace, the efficiency of hot gas cyclone should be improved to reduce the mass flux of fine particles.

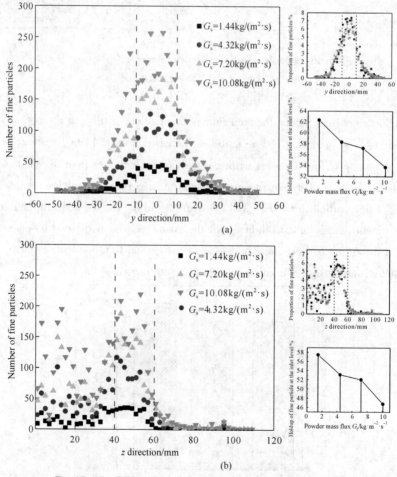

Fig. 15-15 Effect of mass flux of fines on the distribution of hold up of fines in the orifices of the packed bed
(a) y direction (transverse direction); (b) z direction (longitudinal direction)

The pressure distribution along the longitudinal direction under different mass fluxes is shown in Fig. 15-16. It is expected that with the increase of fine mass flux, the number of static fine particles in the packed bed increases. Hence the pressure of the packed bed also increases. When the mass flux increases from 1.44 to 10.08kg/(m²·s), the pressure at $z=60$mm increases from 834 to 998Pa.

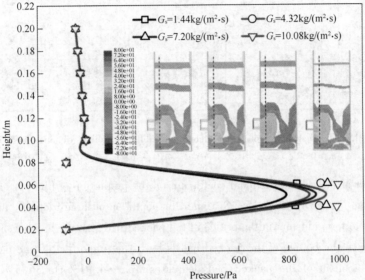

Fig. 15-16 Pressure distribution along with the longitudinal direction under different mass fluxes

15.3.4 Effect of Rolling Friction

Rolling friction provides an effective mechanism to control the translational and rotational motions. Specially, it significantly represents the energy dissipation mechanism with the elastic deformation at the contact point and could act directly on particle dispersion. Hence, the coefficient of rolling friction is one of important impact factors on the clogging behavior of fine particles. Fig. 15-17 shows the distribution of the fine particles at the stable state under different rolling coefficients. As the rolling friction coefficient of fine particles increase, the distribution of dispersed fine particles in the packed bed does not change significantly, while the number and the size of the cluster increase. Typically, at about 20mm from the entrance (about the third layer of the void, the inlet wall is the first layer of void), as shown in the circle in Fig. 15-17, with the increase of rolling friction coefficient, the clusters become more and more apparent, and the size becomes larger. This phenomenon has also been found in previous work. The fine particles enter the inside of the packed bed via gas flow, forming the first accumulation layer in a bottleneck caused by friction between the fine particles and packed particles. The fines then accumulated on the first clogged layer at the bottleneck. The maximum amount of accumulation was determined by the rotational resistance of the particles.

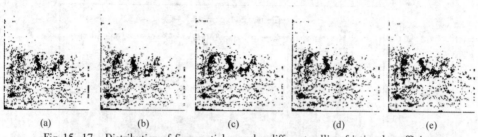

(a)   (b)   (c)   (d)   (e)

Fig. 15-17 Distribution of fine particles under different rolling frictional coefficient
(a) $\mu_r = 0.01$; (b) $\mu_r = 0.02$; (c) $\mu_r = 0.04$; (d) $\mu_r = 0.06$; (e) $\mu_r = 0.08$

The effect of the rolling frictional coefficient on the proportion of fine particles settling at the bottom is shown in Fig. 15-18. Under the condition of large friction, the fines are more likely to form clusters in the orifice of the packed bed, so the number of fine particles settling at the bottom of the packed bed is less. The proportion of fine particles settling at the bottom of the packed bed decreases from 44.91% to 36.3% when the rolling frictional coefficient increases from 0.01 to 0.08.

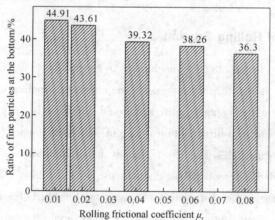

Fig. 15-18 Effect of rolling frictional coefficient on the proportion of fine particles settling at the bottom

Fig. 15-19 shows the distribution of hold up of fine particles in the orifice of the packed bed under different rolling frictional coefficients. The overall fine particles distributions along with the transverse and longitudinal directions under different rolling frictional coefficients are similar to those under other crucial parameters (shown in Fig. 15-11 and Fig. 15-15). At the inlet latitude, the number and ratio of static fine particles in the y and z directions both increase with the increase of rolling friction coefficient. In the actual operation of the COREX process, the fine particles carried by the reducing gas entering the shaft furnace mainly come from the pyrolysis of lump coal, cracking by burden charging and powdering of iron ores with the reduction in melter gasifiers. These fine particles are irregular, and their rolling frictional coefficients would be large, which is one reason that promotes the choking of the slot in shaft furnaces.

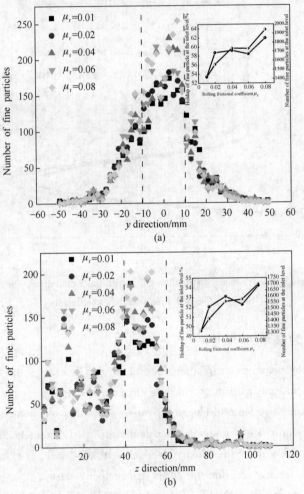

Fig. 15-19 Effect of rolling frictional coefficient on distribution of hold up of fines in the orifice of packed bed

(a) y direction (transverse direction); (b) z direction (longitudinal direction)

Fig. 15-20 shows the effect of rolling frictional coefficient on pressure distribution along with the longitudinal direction. With the increases in the rolling frictional coefficient, the pressure increases. When the rolling frictional coefficient increases from 0.02 to 0.04, the pressure increases from 867Pa to 980Pa, while when the coefficient continues to increase to 0.08, the pressure only increases from 980Pa to 988Pa. This phenomenon is consistent with the phenomenon in Fig. 15-17. The change of the macroscopic clogging of fine particles is not apparent during the change of rolling frictional coefficient from 0.04 to 0.08.

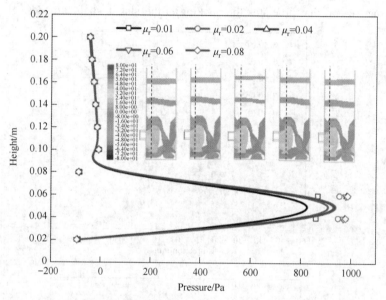

Fig. 15-20 Pressure distribution along with the longitudinal direction under different rolling frictional coefficients

15.3.5 Clogging Mechanism

After clarifying the macro-scale powder distribution and gas flow, the following focuses on the evolution and mechanism of powders clogging in particle-scale. Fig. 15-21(a) presents the compressive force of clogging powders in void space. Some stagnant powders form cluster of certain sizes, and some powders are settled discretely. With the increase of gas velocity, the size of the cluster decreases. In this work, red particles indicate high compressive force, and for the lower gas velocity cases, the red particles are mainly concentrate on the right side of the cluster. However, for the higher gas velocity cases, in addition to the right side of the clusters, there also some discretely dispersed red particles. The evolution of the powders interaction normal force of local cluster that is indicated in Fig. 15-21(a) with time is shown in Fig. 15-21(b). The bottleneck of void spaces in a packed bed is the key point for cluster formation of clogged powders. At $t=0.5s$, some powder particles simultaneously enter the narrow part of the orifice at the same time and they are stuck on the packed particles. These fine particles called 'bottleneck particles' in the rest of the paper. After the bottleneck powders are formed, the following fine particles collide to the preceding bottleneck fines and these collision bring little motion of these stuck fine, and finally these fine particles settle there. These fine particles piled on the bottleneck fines have a small force and show blue in Fig. 15-21(b). Therefore, the amount of powder particles in an orifice directly influence the probability of forming bottleneck particles, thus affecting the fine particles clogging rate.

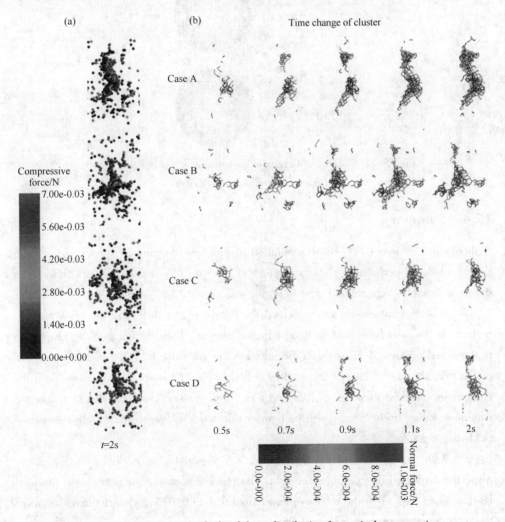

Fig. 15-21 Snapshots of calculated force distribution for vertical cross section

For the cluster in Fig. 15-21, the clogging is mainly due to the interaction between fine particles, creating arches on the packed bed and stopping the flow. When the fine particles form a bridge across the pore throat of the orifice, the bottleneck of void space becomes the starting point for blockage formation. A schematic diagram of the cluster of fine particles in the packed bed can be explained in Fig. 15-22. For the discretely clogging fine particles, the mechanism is mainly due to the drag force and friction caused by a small fine particle rolling on the surface of large particles whose spacing is close to the diameter of fine particles.

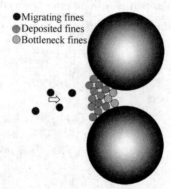

Fig. 15-22 Schematic diagram of fines migration and bridging in the packed bed with lateral injection

15.4 Summary

A three-dimensional CFD-DEM simulation of powders clogging in a packed bed with lateral inlet was performed. The distribution of hold up of powders and gas flow under different variables was studied. The following conclusions are obtained:

(1) With the increase of gas velocity, the more powders diffuse in the transverse direction. At the inlet level zone in longitudinal direction, both the number and proportion of deposited powders decrease with the increase the gas velocity. The gas distribution injected from the lateral inlet of the packed bed shows a 'J' shape for all cases. With the continuous injection of powders, the pressure drop gradually increased and reached a maximum value. In the period with no powder injected, the pressure drop decreases and tends to be stable.

(2) With the decrease of diameter ratio, the cluster in the packed bed decreases, while the dispersed fine particles increase. The number of static fine particles at the inlet latitude decreases when the d/D decreases from 0.1 to 0.075, while it increases when the d/D decreases from 0.075 to 0.05.

(3) The number and size of cluster increase significantly as the mass flux of fine particles increases. With the rise of mass flux, the number of static fine particles in the orifice of the packed bed at the inlet latitude gradually increases.

(4) With the increase of the rolling frictional coefficient, the number and size of the cluster increase. The number and ratio of static fine particles in the y and z directions at inlet latitude both increase with the rise of the rolling frictional coefficient. The fine particles carried by reducing gas entering the COREX shaft furnace are mainly irregular, which is one reason that promotes the choking of the slot in shaft furnaces.

(5) Two kinds of clogging powders inside the porous. One is a cluster composed of agglomerated powders, the other is a dispersed single powder. The cluster is mainly due to the interaction between fine particles, creating arches on the packed bed and stopping the flow. When the fine particles form a bridge across the pore throat of the orifice, the bottleneck of void space becomes the starting point for blockage formation.

References

[1] Cao S C, Jang J, Jung J, et al. 2D micromodel study of clogging behavior of fne-grained particles associated with gas hydrate production in NGHP-02 gas hydrate reservoir sediments [J]. Mar. Petrol. Geol, 2019, 108: 714~730.

[2] Khilar K C, Fogler H S. Migrations of fines in porous media [M]. Boston: Kluwer Academic Publishers, 1998.